PRACTICAL

PLANT PHYSIOLOGY

BY

FREDERICK KEEBLE, Sc.D.

PROFESSOR OF BOTANY AND DEAN OF THE FACULTY OF SCIENCE
UNIVERSITY COLLEGE, READING
AUTHOR OF "PLANT-ANIMALS"

ASSISTED BY

M. C. RAYNER, B.Sc.

LECTURER IN BOTANY IN UNIVERSITY COLLEGE, READING

British Library Cataloguing-in-Publication Data
A catalogue record for this book is available from
the British Library

Botany

The term 'botany' comes from the Ancient Greek word *botanē*, meaning 'pasture', 'grass', or 'fodder', in turn derived from *boskein*, meaning 'to feed or graze'. It chiefly involves the study of plant life, as a branch of biology.

Traditionally, botany has also included the study of fungi and algae by mycologists and phycologists respectively, with the study of these three groups of organisms remaining within the sphere of interest of the International Botanical Congress. Nowadays, botanists study approximately 400,000 species of living organisms of which some 260,000 species are vascular plants and about 248,000 are flowering plants.

Botany originated in prehistory as herbalism with the efforts of early humans to identify – and later cultivate – edible, medicinal and poisonous plants, making it one of the oldest branches of science. Examples of early botanical works have been found in ancient texts from India dating back to before 1100 BCE, in archaic Avestan writings (an Iranian language known only from its use in Zoroastrian scriptures), and in works from China before it was unified in 221 BCE.

Modern botany traces its roots back to Ancient Greece, specifically to Theophrastus (c. 371–287 BCE), a student of Aristotle who invented and described many of its principles. Today, he is widely regarded in the scientific community as the 'Father of Botany'. Theophrastus's major works, *Enquiry into Plants* and *On the Causes of Plants* (both looking at plant structure, variety, reproduction and growth), constitute the most important contributions to botanical science until the Middle Ages, almost seventeen centuries later. Another work from Ancient Greece that made an early impact on botany is *De Materia Medica*; a five-volume encyclopaedia about herbal medicine written in the middle of the first century by Greek physician and pharmacologist Pedanius Dioscorides. *De Materia Medica* was widely read for more than 1,500 years subsequently.

Medieval physic gardens, often attached to monasteries, contained plants of great medicinal importance. They were forerunners of the first botanical gardens attached to universities, founded from the 1540s onwards. In the mid-sixteenth century, botanical gardens were founded in a number of Italian universities – and the Padua botanical garden in 1545 is the first of such, still in its original location. These gardens continued the practical value of earlier 'physic gardens' of the monasteries, and further supported the growth of botany as an academic subject.

Botanic gardens encouraged the work of academics such as the German physician Leonhart Fuchs (1501–1566). Fuchs was one of 'the three German fathers of botany', along with theologian Otto Brunfels (1489–1534) and physician Hieronymous Bock (1498–1554). Fuchs and Brunfels broke away from the tradition of copying earlier works to make original observations of their own, whilst Bock created his own system of plant classification. In 1665, using an early microscope, another famed botanist, the Polymath Robert

Hooke (1635 - 1703) discovered 'cells' in plant tissue, a term he coined. During this early period, lectures were also given about the plants grown in the specially constructed botanic gardens, and their medical uses demonstrated. Botanical gardens came much later to northern Europe – largely due to the obvious differences in temperature. The first in England was the University of Oxford Botanic Garden, constructed in 1621.

Efforts to catalogue and describe the collections of these gardens were the beginnings of plant taxonomy, and led in 1753 to the binomial system of Carl Linnaeus that remains in use to this day. Linnaeus's system was a hierarchical classification of plant species providing a solid reference point for modern botanical nomenclature. This established a standardised binomial or two-part naming scheme where the first name represented the genus and the second identified the species within the genus. For the purposes of identification, Linnaeus's *Systema Sexuale* classified plants into twenty-four groups according to the number of their male sexual organs. The twenty-fourth group, *Cryptogamia*, included all plants with concealed reproductive parts, mosses, liverworts, ferns, algae and fungi.

Increasing knowledge of plant anatomy, morphology and life cycles led to the realisation that there were more natural affinities between plants than Linnaeus had indicated however. Scholars such as Adanson (1763), De Jussieu (1789), and Candolle (1819), all proposed various alternative natural systems of classification that grouped plants using a wider range of shared characters and were extensively followed. The Candollean system reflected his ideas of the progression of morphological complexity and were developed the classifications by Bentham and Hooker, influential until the mid-nineteenth century. Darwin's publication of the *Origin of Species* in 1859 and his concept of common descent, further required modifications to the Candollean system to reflect evolutionary relationships as distinct from mere morphological similarity.

In the nineteenth and twentieth centuries, new techniques were developed for the study of plants, including methods of optical microscopy and live cell imaging, electron microscopy, analysis of chromosome number, plant chemistry and the structure and function of enzymes and other proteins. In the last two decades of the twentieth century, botanists exploited the techniques of molecular genetic analysis, including genomics and proteomics and DNA sequences to classify plants more accurately. Particularly since the mid-1960s there have been advances in understanding of the physics of plant physiological processes such as transpiration (the transport of water within plant tissues), the temperature dependence of rates of water evaporation from the leaf surface and the molecular diffusion of water vapour and carbon dioxide through stomatal apertures.

Twentieth century developments in plant biochemistry have been driven by modern techniques of organic chemical analysis, such as spectroscopy, chromatopgraphy and electrophoresis. With the rise of the related molecular-scale biological approaches, the relationship between the plant genome and most aspects of the biochemistry, physiology, morphology and behaviour of plants can be subjected to detailed experimental analysis. Such developments have enabled advances in areas as diverse as pesticides, antibiotics and pharmaceuticals, as well as the practical application of genetically modified crops designed for traits such as improved yield.

The study of botany is an incredibly important science, as plants underpin almost all life on earth. They generate a large proportion of the oxygen and food that provide humans and other organisms with aerobic respiration with the chemical energy they need to exist. In addition, they are influential in the global carbon and water cycles and plant roots bind and stabilise soils, preventing soil erosion. Plants and the science of botany are crucial to the future of human society – allowing insight into our food, natural environment, medicine and products. It is a branch of human endeavour with an incredibly long and varied history, and it is hoped the current reader enjoys this book on the subject.

SAND CULTURES.—POPPIES GROWING IN STERILE SAND.

A. Watered with normal culture solution.   B. Watered with a solution lacking nitrates.   (See Chapter VIII.)   (From a photograph.)

PREFACE

THE purpose of this book is to provide students and teachers with an outline of the experimental investigations on which our knowledge of the physiology of plants is based.

It would have been easier to present the most attractive of the facts and theories of this branch of science than to make the attempt, of which this book is the outcome, to provide a practical text-book which should serve as an educational instrument and as a stimulus to independent observations.

The student, however, cannot grasp the facts of Natural Science by reading books on the subject. Though the man of mature mind, trained in the art of sifting and grouping facts, and in the habit of seizing essential truths, may gain a useful acquaintance with Natural Science by the perusal of text-books, the student cannot. He needs a method of education which serves, not only to furnish, but also to train the mind. In other words, the student requires both information and discipline. He should be told less, and find out more. It cannot be disputed that the element of intellectual discipline is ignored too much by modern educational methods, nor that, as a result, students suffer from lack of training.

In science, in particular, the teacher is too much concerned with rendering the subject attractive and too little with making it to serve the end of developing the reasoning and imaginative faculties of his pupils.

The method of teaching Natural Science by means of lectures and a ritual of practical work associated therewith is fundamentally wrong. It appeals to the memory but sterilizes the imagination. It preaches comfortable things to the student, who ought, instead of being submitted to scientific sermons, to be experiencing the process, at once pleasurable and painful, of teaching himself.

In the course of the lecture hour, the teacher covers a large area of ground which is a terra incognita to the student. From the fulness of his experience, the lecturer clears the ground of all difficulties which beset the path of the scholar, with the result that the latter, though he may, or may not, have enjoyed the personally-conducted tour, has learned nothing of the art of exploration. The student would be unable, indeed, to find his way again over the same ground if left to take the journey alone.

Subjected to this lecture-method of instruction, the student becomes the repository of the second-hand. What he is told to look for, he sees with the teacher's eye. If the method is successful and the student docile, the latter passes his examination with distinction, but remains in ignorance of the importance of initiative. The less docile student takes and expresses—often by idleness—a strong dislike to science, preferring, instinctively, some subject more calculated to train the mind or less certain to overload the memory.

In either case, the lecture-method, which dates from the time when books were scarce, or dear, or bad, accomplishes little in comparison with what might be achieved if less ground were traversed during the lecture-excursion and if the student, instead of being told things, was encouraged to observe and think for himself.

Until the present, deplorable method is replaced by a better, Natural Science, which must come in time to be an essential part of all educational curricula, must submit to the reproach of being of less value, as a subject of mental training, than Mathematics or Classics. The reproach is not deserved. Without a knowledge of Natural Science, men and women remain blind to half the beauty and meaning of life : without it, they are ignorant of the modes whereby all manner of pressing problems, social, political and other, ought to be attacked if they are to be solved.

If reform is to be effected in the teaching of Natural

Science, it must be by making the subject less an affair of information and more a matter for thought. The logical argument which runs through the Natural Sciences, cohering the innumerable facts to a whole, must be displayed, and the student must find, in his scientific studies, means of developing the higher functions of his mind, namely, the powers of thinking rightly and imagining freely.

The proper method of teaching Natural Science is as follows : Very young children should be rendered familiar with large numbers of natural objects, including, of course, plants and animals. The purpose of this preliminary course is to make the child take cognisance of natural objects and phenomena; of flowers and their names, of fruits and seeds, of buds and leaves, of herbs and trees. In short, confining our remarks to the biological side of this course, the child should learn by means of it to see and watch and care for some of the myriad living things of field, hedgerow and garden.

This is the Nature Study course, in which no attempt should be made to force explanations or to develop the reasoning faculty of the child; the object being to furnish the mind.

Having achieved this end, Nature Study has exhausted its educational service. It is as useless in later stages as it is valuable in the early stage of education.

The Nature Study course should be succeeded in the school by one on elementary physics and chemistry. Such a course is, as a fact, followed in many schools. In it, the properties of matter are first investigated experimentally, and, later in the course, the relations of facts or phenomena, one with another, are studied. During this course, the child begins to reason and to imagine scientifically.

Then should follow the third part of the course in Natural Science, which should be of a biological nature.

It is a great misfortune that this third part but rarely finds a place in the school curriculum. The scholars continue too long with the more mechanical Natural Sciences, Physics and Chemistry, and leave school only too often with no knowledge of Biology.

This is the case generally with boys, though, not infrequently, girls follow a course of morphological Botany instead of one on Physics and Chemistry.

In either case, the results are bad. Neither the boys nor the girls know anything of the modes of life of plants or animals; the boys, because they have not studied the subject; the girls, because, without a knowledge of the elements of Physics and Chemistry, they cannot understand it. The remedy is to require the scholars, boys and girls alike, to follow the elementary course in Physics and Chemistry, and then to proceed to one on the elements of Biology.

The object of this book is to provide such a biological course; that is, one suitable for the higher classes in schools and for the first year class in the University.

It is true that the book deals only with the physiology of plants; but it is true also that anyone who is familiar with the facts of this science knows not a little of the essentials of animal-physiology.

By following the course, the training in reasoning and in imagination, begun during the study of Physics and Chemistry, is extended. The scholar who has passed through the three courses outlined above possesses a confident knowledge of the most important facts of Natural Science. He has also a trained mind.

This book is useless as a reading-book; but, if its intention has been fulfilled, it should prove a serviceable tool wherewith the student may dig out for himself the fundamental truths of the science.

To do this means, of course, hard work on his part; but once the student has learned that one of the chief pleasures of life lies in hard, intellectual work—just as another great pleasure lies in the hard, physical work which he bestows on his games—he will, it is to be hoped, prefer this course to the easier, sitting-down method wherein he plays the part of a recipient of knowledge of which he knows neither the origin nor the mode of getting.

If the book is to serve the end for which it has been designed, it must be used, like any other tool, in a proper way.

The teacher should decide on the number of chapters which are to be dealt with in any special course. He should then divide the chapters into lessons, each of which should be accomplished in the hours which may be devoted to the subject in the course of the week. Having explained that the need for making one experiment follows from the conclusions drawn from its predecessor, the teacher should point out to the students that they should pay no attention to the results of any experiment recorded in the text-book until they themselves have obtained records.

Before starting an experiment, the student should read the instructions both in the text and in the appendix, so that he may know all the details necessary for the setting-up and use of the apparatus. Having done this, the student should make, in each case in which the experiment admits of it, a working drawing of the apparatus which he proposes to construct, and he should write a brief prediction of the results which he expects to obtain by the performance of the experiment. When the apparatus is made, and before he begins to use it, the student should examine it critically with the object of making sure in advance that the apparatus can do the work required of it. Should the apparatus fail to work properly, the student must submit it again to critical examination in order to discover the cause of the failure.

Nothing is more valuable for the education of students than learning to construct apparatus, making it work, discovering the reasons of its failure and devising means for overcoming its defects.

Instead of expending his energies in lecturing, the teacher should devote himself to obtaining material suitable for the experiments, assisting students in overcoming difficulties and securing permanent records of the experiments. These records should be deposited in the physiological museum, to which frequent reference is made in the text.

When a lesson is completed, the teacher should discuss the results and conclusions with the students, and point out the bearing of these results on other parts of the subject.

In conclusion, though the author must be held responsible for any errors which are contained in the book, and for the opinions expressed in the preface, he acknowledges with pleasure, what is acknowledged already in the title-page, that, in writing this book, he has had the advantage of the assistance of his colleague, Miss M. C. Rayner.

Nearly all of the illustrations have been made expressly for this volume. Some are reproductions from photographs, others are due to the skill and kindness of Miss Dorothea Richardson, from whose original drawings they have been reproduced.

FREDERICK KEEBLE.

University College, Reading,
December, 1910.

CONTENTS

CHAPTER I

The problems of plant-physiology and the method by which they are to be solved. The scientific method. Classification of physiological problems.

CHAPTER II

The mode of germination of seeds: the parts of the seed and seedling: the resting and active states of seeds: the resisting powers of resting seeds: germination capacity: the visible order of events in germination. The nature and function of cotyledons and of endosperm: adaptation in plants: large seeds and small seeds.

CHAPTER III

The nature and chemical properties of the food-substances contained in the cotyledons and endosperm of seeds.

CHAPTER IV

The changes undergone by the reserve food-materials of the seed during germination: the mode of passage of food-

CONTENTS

materials from the place of storage (endosperm or cotyledons) to the place of consumption (the growing embryo).

CHAPTER V

The meaning of the term Nutrition : the use which the plant makes of food-substances. The germinating seed considered as a machine. The source of the power which drives the machine and the conditions under which it works.

CHAPTER VI

The seedling as an independent plant : the lowest forms of plants and animals and the lines followed in the evolution of the higher plants and animals. The distinguishing characters of root- and shoot-systems. The mode of growth of the root : the functions of its parts : the root-hairs, the absorbent organs of the root.

CHAPTER VII

The way in which water is absorbed by root-hairs and other cells. Osmosis and osmotic pressure. The plant-cell as an osmotic apparatus.

CHAPTER VIII

The substances taken up by the roots of plants. The composition of plant-ash. Water- and sand-cultures. The soil in relation to plant-life. The origin of soils : their physical, chemical, and biological properties.

CHAPTER IX

The absorption and loss of water by the plant. The water-requirements of various types of plants :—hygrophytes and xerophytes. The process of the transpiration of water by the leaves : the structure of the leaf in relation to this process : the part played by stomata : the opening and closing of stomata and the conditions under which these movements occur. Apparatus for measuring rate of transpiration—(Potometer).

CHAPTER X

The passage of water from root to leaves : the channels followed by the transpiration current : water-conducting wood and skeletal wood. The causes of the ascent of water. Phenomena connected with the absorption of water : root pressure : bleeding : excretion of water : water-pores (hydathodes).

CHAPTER XI

The origin of the carbon compounds contained in plants. The raw materials from which the plant constructs these compounds. The part played by chlorophyll grains (chloroplasts) in the manufacturing process : the energy by which the process is carried on. The passage of carbohydrates from the leaves to other parts of the plant. The synthesis of organic nitrogen compounds in the plant.

CHAPTER XII

The modes of response of plants to stimulation. Irritability. The reflex-actions of plants. Tropisms (Geotropism, Photo-

CONTENTS

tropism, etc.). Morphogenetic responses. The component parts of a reflex-action :—perception, excitation, transmission of nervous impulses, excitation and response of the motor region.

CHAPTER I.

A SEED sown under suitable conditions germinates, giving rise to a seedling. The seedling grows, puts forth leaves and branches, drives its roots further and further into the soil, and becomes a mature plant. Presently and in due season the flowers appear, endure in their delicate beauty for a while, and, their work being done, fade away. The stalks which bore the flowers now support the swelling fruits within which the seeds are ripening. When the seeds are set, the fruit bursts open, scattering them far and wide, or, falling to the ground and rotting, sows the seeds near the parent plant. Such are the more striking episodes in the life of a flowering plant.

This regular sequence of events—the germination of the sown seed, the formation of roots and branches, leaves and flowers, the setting of the seeds and the ripening of the fruits—seems so natural that we are apt to accept it as needing no explanation. But when we begin to observe the several processes more closely, to reflect upon them, or to compare one plant with another with respect to them, we find ourselves asking all sorts of questions. Why is it that the chickweed of the hedgerow runs its course from seedling to fruiting stage in less than one brief season, whilst a foxglove does not reach the flowering stage till its second year? By what gymnastic exercises does the seedling extricate itself so neatly from its seed coat? How is it that the plantain on the lawn hugs the ground so closely as to escape almost uninjured the knives of the cutting machine?

Whence come the power and the material by means of which the giant oak with gnarled trunk and spreading branches forms itself from an acorn? How is it that, no matter which way up we plant a bean seed in the soil, the root of the seedling turns downward and burrows in the earth, whilst the stem twists itself so that it comes to grow upward into the air? From what source does the apple obtain the sugar to which it owes its sweetness or the rose the perfume with which it scents the air?

When once we begin to interest ourselves in plants, we find that the problems which they suggest are as varied as they are numerous, and we realise that in every plant of hedgerow, field, or garden, all sorts of strange events are happening.

Could we but find answers to the questions which living plants suggest to us, we should be in possession of a great body of knowledge concerning their life and work; in other words, we should have taught ourselves not a little of the science of plant physiology.

Hence the most pressing of our problems is, how are we to set about obtaining answers to any of the questions which arise in our minds when we observe living plants? Curiosity suggests the problems, how does science seek to solve them?

In order to find out the method of scientific discovery, let us fix our attention on some particular phenomenon exhibited by a plant, and consider how we may ascertain its significance.

The phenomenon which we will choose for investigation is the origin of the drops of water which appear on the leaves of certain plants, such, for example, as the oat.

1.† We sow oats in ordinary soil in two pots, and, when the leaves of the seedlings are four or five inches in height, we may find, on examining them in the early morning, that near the apex of each leaf is a shining drop of water looking like a dew-drop (Fig. 1).

We want to discover whence the water-drops come !

Though we were to sit up all night watching the plants, we should obtain no solution of the problem : all we should

† The numerals, in heavy type, which occur throughout the book refer to the experiments which are to be performed, see Preface, p. **xi.**

see is that the drops, when first formed, are small, and that they may increase in size very rapidly. Observation, therefore, though it provides us with scientific puzzles, does not

FIG. 1.—OAT SEEDLINGS.
Drops of water (w), excreted from water pores (hydathodes) in the tips of the leaves.
From a Photograph.

necessarily help us to solve them. Observation, careful and continued observation of the living plant, is essential for the study of plant physiology, but something besides observation is wanted now.

Evidently all we can do is to make a guess as to the origin of the water-drops. Confronted with the problem, we call imagination to our assistance, and by its aid guess at the answer. As the result of guessing, we suggest that the water may be dew formed from water-vapour in the air. Next we ask ourselves; suppose the guess is wrong, what then? At once common-sense makes answer; if the water does not come from the air, it must come from the plant itself. Having exhausted our guessing powers, we proceed to look coldly at the alternative suggestions, and to review them in the light of common-sense. In this case, common-sense admits that either guess may be right. But our guesses cannot both be right. Therefore we must discover some way of deciding between them. If the problem were one of a kind with which we are more familiar, for example, as to the height of a friend; and if two people guessed differently with respect to this, we should not hesitate as to our method of verification. We should stand the friend against the wall and measure him. That is, we should put the guesses to the test of experiment. Whereas no amount of discussion would determine the correctness of the guesses, a yard measure properly used would settle the matter in a minute.

In like manner, to solve the problem of the origin of the water-drops, we submit it to the test of experiment. But how is this to be done? Once again we must appeal to imagination and common-sense. We must use these faculties conjointly in order to invent an experimental test. In our particular problem, it is easy enough to devise a method. We know that plants take up water from the soil, and so we argue thus : if the drops of water on the leaves come from the air, they may make their appearance as readily on the unwatered as on the watered plants; but if the drops come from the plants themselves, it will probably matter fundamentally whether the plants contain much or little water. Thus we arrive at our method of experiment. Water one pot thoroughly, withhold water from the other, and examine the plants on successive mornings. When we do this, we find that the water-drops are plentiful on the leaves of the watered seedlings, and are absent from, or, at all events, fewer on the others.

To complete the proof we devise a further experiment.

2. For example, give water to the previously unwatered pot, stand it under a bell jar, and observe that in five or ten minutes drops of water appear on the tips of the leaves. Therefore we conclude that the water which appears on the leaves comes from the plant and not from the air.

But in solving the particular problem, we have discovered also the answer to the general problem—how to set about getting replies to special questions? The answer is—by using exactly the same method as that which we have just employed. There is indeed no other way. It is called the scientific method, and involves, as we have learned, processes of guessing, reasoning, and trying.

Thus the processes involved in the use of the method are as follows:

1. The guessing process in which the imagination is invited to suggest possible answers to the problem under investigation. Our guesses may be as wild as we like to make them. The more the imagination is allowed to run riot, the more likely are we to open up new paths for investigation. Indeed, it is no exaggeration to say that the greatest discoveries are the outcome of the wildest guesses.

2. The judging process, in which common-sense assumes the part of advisor, recommending this or that guess as more likely to prove true, and rejecting any guess which runs counter to established truth.

3. The testing process, which consists in the devising and execution of experiments calculated to demonstrate the truth or falsity of the guess, or, as it may be called, the hypothesis, which gains the approval of common-sense.

4. The summing-up process, by which we decide, whether the evidence provided by the results of the experiments is convincing or not.

If the evidence is absolutely conclusive in favour of our hypothesis, we speak of that hypothesis as a fact; if the evidence is inconclusive, we may yet continue, for want of a better, to hold the hypothesis and to use it in our arguments; though in doing this we have to be extremely cautious, and to remember that our hypothesis is "not proven."

Hence to study a science aright is not to become a narrow

specialist, but to develop all the highest faculties of the mind. This scientific method is not peculiar to plant-physiology : it is the method employed in all the sciences, and by its use all the knowledge of nature which we possess has been obtained.

All that remains to be done in this introductory chapter is to classify our problems, that is, to arrange those of like nature in groups, and the groups in convenient order. In doing this we will make an assumption, which may not, at first sight, seem very probable; but which will be of great service to us. Whether the assumption is true or false we shall discover as we proceed with our investigations. We assume that the life of a plant is not different in essentials from that of an animal or from that of man himself. Unless this assumption is wholly false, and, in that case, we shall soon discover our mistake, it will be of great assistance to us in the otherwise puzzling problem of the arrangement of our questions. For we know, without the aid of science, and from our common experience, a good deal about our own life-processes. We know, for example, that we feed, and that without food of certain kinds we cannot live. We know that we breathe, and that we cannot exist for more than a few minutes without air. We know also that we, like animals in general, move, and that some movements, as, for example, getting up in the morning, depend on an effort of will, whilst other movements, for example, the beating of the heart, are independent of consciousness. We know also that animals and plants grow, give birth to young, and ultimately die.

Hence we arrive at the following classification of the problems of plant physiology :

(1) Feeding processes (nutrition).
(2) Breathing processes (respiration).
(3) Growth processes.
(4) Phenomena of movement and of sensitiveness (or irritability).

CHAPTER II.

A FULLY grown plant is by no means a convenient subject for experimental purposes. Not only is it bulky, but its roots are hidden in the ground and cannot be disturbed without damage to the plant. On the other hand, a handful of pea or bean seeds may be obtained for a penny, and, when planted, the seeds produce seedlings in the course of a week or two. Moreover, inasmuch as the seedlings grow rapidly, we may assume, from analogy with young children, that they are likely to feed hungrily. Hence seeds and seedlings should prove very useful to us in our studies in plant-nutrition. We will therefore commence our work by an examination of seeds and seedlings.

Since we shall require seeds for all sorts of experiments, we must take every opportunity of getting together a large and varied collection. At the proper times of the year, ripe seeds of garden plants, weeds, and common trees should be gathered, dried, and stored in corked or stoppered bottles. The bottles should be labelled, and on each label should be written the name of the seed (or fruit), the locality whence it was obtained, and the date of gathering. If it is not possible to collect a sufficiently varied assortment of seeds, some may be purchased from seedsmen and stored in labelled bottles. Samples of the seeds and fruits should be affixed to cards with the name, natural order, and other interesting details, such, for example, as locality and weight, and the cards placed in the physiological museum, in which records of experiments, specimens, and photographs, etc., should be kept. In case the beginner does not know how to distinguish seeds from fruits—and some fruits look exactly like seeds —he should refer to an elementary text-book (Bibliography, 3, 5), which deals more particularly with the morphology of plants, that is, with the characters and peculiarities of their form and structure. For, though we, in studying the work of plants—that is their functions— shall have to take notice of their form and structure, we have not space to deal fully with the morphological branch of botany.

Having become familiar with the shapes, sizes, and peculiarities of the seeds and fruits of the commoner plants, we proceed to germinate some peas. At once the question arises : since the seeds in our collection do not germinate whilst in the bottles, what is to be done in order to make them begin to grow? Now everyone who has access to a garden or to the country knows how quickly weeds and other plants spring up in showery weather, and hence we make the sure guess that a supply of water is necessary for germination. Even though we know this, we prove it; for, by so doing, we shall extend our knowledge and make it more precise.

3. To this end, prepare three pots of garden soil, dry one thoroughly in a kitchen oven, and, in order to prevent the soil from getting moist again, set it to cool under an inverted marmalade jar, or similar vessel. See that the soil in the other two pots is thoroughly moist. Determine the average weight of ordinary, dry pea seeds by weighing several samples of twelve each. Put a couple of dozen seeds to soak in tepid water, and, after twenty-four hours, dry their surfaces by means of a cloth, weigh and compare them with respect to weight, size, and shape with the dry seeds. Calculate the percentage of water taken up. Now plant four or six peas in each of the three pots : putting dry seeds in the pot with the dry soil, dry seeds in one of the pots with moist soil, and soaked seeds in the remaining pot. Label the pots 1, 2, 3, and note in a rough note-book the details of time of planting, and states of seed and soil. Cover each of the pots with a glass plate or piece of cardboard or stiff brown paper, and see that the soil in pots 2 and 3 does not get dry. Record the dates of appearance of the seedlings in each of the pots. Copy out the results neatly in a note-book kept for the purpose, and add any remarks that seem interesting. Records should be made of the results of every experiment that is performed, and, whenever it is useful, sketches should accompany the records, which should be arranged in brief, tabulated form.

4. Take the remaining soaked seeds, wipe them, put them in a dry place—for instance, on a shelf in a living room—and weigh them at daily intervals, and thus determine the rate at which they lose water. When they seem fairly dry, put them in a thin paper bag in a desiccator. (See Appendix A.) At the same time, weigh and place in a paper bag a dozen dry, unsoaked peas, and put this bag, properly labelled, in another desiccator. After an interval of about a week, weigh the two lots and determine how much each has lost in weight. Leave them exposed to the air of a room for some hours, weigh them again, and compare these weights with those of the same seeds when taken from the desiccator. From the experiments, it is evident that seeds are hygroscopic, that is, they take up water from moist air and give up a certain amount of water when the air to which they are exposed is dry. The bearing which these facts have on such matters as the following should be considered :—the importance of storing seeds out of contact with moist air : the difference that the weather at the time of harvesting is likely to make to the viability of the seed : the advantage and possible disadvantage of soaking bean or pea seeds before sowing in the garden : the fact that, in wet autumns, seeds of various plants may be found germinating whilst still attached to the parent plant.

In cases where students work in groups, some should use one kind of seed for the above experiments and others another, e.g. barley grains (which are strictly fruits), horse chestnuts, onion seeds, etc. The results obtained with these different seeds should be compared with one another.

We have now confirmed our knowledge that seeds,

in order to germinate, require water; we have found that the amount of water which seeds, such as peas, can absorb is surprisingly large; and we have learnt also that seeds are hygroscopic. We recognise that a knowledge of these facts helps us to store our seeds properly, and shows us how we may hasten their germination. We will next determine whether seeds dried as thoroughly as possible in a desiccator are absolutely dry, or whether they still contain water.

5. To this end weigh a dozen peas which have been in the desiccator for a week, soak them till they are soft, wipe them with a cloth, and pound them in a mortar; transfer the whole of the mash to a weighed, dry porcelain dish, and dry it thoroughly in a drying oven at about 100° C. After two days, take the dish out of the oven, stand it in a desiccator to cool, and then weigh it. Replace the dish in the oven and continue the weighing at daily intervals till no further loss of weight is recorded. We thus obtain the *dry weight* of the substance of the seeds, and a comparison of this weight with that of the desiccator-dried seeds tells us how much water the latter, apparently dry seeds really contained. The result of the experiment proves that even the driest seed contains a considerable percentage of water. The above experiment will be the more instructive if, at the same time, other vegetable tissues, *e.g.* carrots, turnips, and also fresh leaves (grass or spinach, etc.) are weighed, dried in a desiccator, weighed again, then chopped up (there will be no need to soak them first, as they are not so flinty hard as the seeds), pounded in a mortar, dried in the drying oven, and their dry weights determined.

6. A ready way of proving that ordinary air-dry seeds contain a considerable amount of water is as follows : half fill a wide-mouthed glass bottle with peas. Stopper the bottle, and place it in the drying oven at about 90-100° C. After 1-2 hours remove the bottle and observe that, as it cools, water, given off by the peas, condenses to form drops on the sides. By using different kinds of vegetable tissue it will be discovered that they all contain a certain amount—some a very large amount—of water, and that, of vegetable structures, seeds contain far less water than any others. That it is to this fact that seeds owe their resistant powers we demonstrate in the following way.

7. Prepare a saucepan of boiling water, place a few ordinary dry peas and equal numbers of soaked and of desiccator-dry peas in small canvas or muslin bags, plunge them in the boiling water for a few seconds, plant the three lots (after soaking the dry seeds) in pots, and record their germination. Whereas the thoroughly dry seeds have not been injured by their short immersion in boiling water, the soaked seeds show by their failure to germinate that they have been killed. It is noteworthy that advantage is taken of the resistance of dry seeds to high temperatures in treating grains of oats, the surfaces of which are suspected to be contaminated with the spores of a disease-producing fungus called smut. The grains are plunged for five minutes in water at a temperature of 55° C., and subsequently sown. The effect of the hot water is to destroy the smut spores without injuring the oats.

8. Next, the effects of low temperatures on very dry and on soaked seeds should be determined. The most convenient way to do this is to pound up ice and salt and to put the freezing mixture into a small pail, in the middle of which a glass flask is placed. The several small lots of peas, each lot in a muslin bag, are put into the glass vessel and left there for some hours. The bags of seeds are then withdrawn, and the germination capacities of the three lots of seeds tested.

From the result of this and similar experiments it is learned that dry seeds are more resistant to adverse conditions than soaked seeds.

During the summer we make a comparison between unripe and ripe peas. Definite experiment is not necessary to convince us that the unripe seeds in their young pods contain far more water than the ripe seeds, and we may take it that, during ripening, one process which goes on is the gradual loss of water by the seeds. When, on the one hand, we call to mind the extremes of temperature to which the seeds of plants are exposed during their long winter's sojourn in the ground, and when, on the other hand, we realise the great resistant power of dry seeds, we cannot doubt that this natural drying process, which takes place during the ripening of seeds, is of advantage to the plant, making undoubtedly in many cases the difference between destruction and survival. That the resting state is due, in large measure, to the natural drying during ripening may be inferred from the experiments we have made, and also from the fact already noted that, in wet autumns, various kinds of plants may be met with the seeds of which are already beginning to germinate on the parent plants. Specimens illustrating this phenomenon should be collected and added to the museum.

Experiments recently made have proved that certain seeds may retain their capacity for germination for a great number of years; in one instance, among seeds known to have been kept for 87 years, some were found to be capable of germination, and it is interesting to know that experiments are now in progress to determine for how long seeds, which have been dried as thoroughly as possible, will retain their vitality. Though, as we have just learned, dry seeds may survive for many years, there is no evidence to prove the truth of the statements which are often made that wheat grains and seeds of other plants deposited a thousand or more years ago with mummies in mummy cases in Egypt have retained till the present day their powers of germination. Indeed, there is good reason to believe that such "mummy wheat" has long ago lost its vitality.

Our experiments demonstrate that a seed is a structure which, by reason of its dryness, is capable of passing through a long resting stage. Whilst in the dry state, it is far more resistant than is the growing plant. By providing it with water, the seed may be caused to pass from its resting or latent state into one of activity. What is true of seeds is also true of the simpler reproductive bodies of many of the lower plants. For instance, it has been shown that the resting spores of some bacteria are not destroyed by exposure to such high temperatures as 100°-120° C., at which temperatures the bacteria in their active, growing state are killed. That this great power of heat-resistance is due to the dryness of the spores is proved by the fact that, if the spores are brought under such conditions of moisture and warmth that they begin to grow, they lose their resistant powers. Since the group of plants known as bacteria includes many disease-producing forms, and since some of the latter produce resting-spores, the bacteriologist and the doctor have to take the resistant powers of spores into account in their efforts to exterminate disease-producing germs.

Let us now return to our study of seeds, and set ourselves to find out what is the first visible sign of germination.

9. In order to do this sow samples of soaked seeds, some in earth, others in germinators. Germinators of various patterns may be obtained ready-made (Appendix B), but one of the simplest and most useful may be made from a couple of ordinary saucers. Several layers of thick white blotting-paper are moistened thoroughly and fitted neatly into one saucer. A few soaked seeds are distributed on the blotting-paper, and the other saucer, into which moist blotting-paper may also be fitted, is inverted over them. The only precautions necessary are that the blotting-paper should not be too wet to begin with nor become too dry. The saucers may be covered with a large jar or box and stood in a warm place. Every second day, two or three soaked peas are put into the germinator in order that we may obtain all the various stages of germination at one and the same time. In the course of a day or two, a small white conical structure is to be seen projecting from the first sown seeds. It elongates to form a cylindrical body with a rounded end. By comparing older with younger stages, it will be seen that this body or, at all events, the greater part of it is the young root or radicle. This method of making comparisons backward, that is, of comparing an older with a younger structure, is very useful, and will often enable us to determine the nature of doubtful structures. Whilst the pea seeds are germinating, drawings should be made of each stage up to the time when the various parts of the seedling are recognisable. Germinate also a number of other seeds, *e.g.* mustard or cabbage, onion, etc., in order to demonstrate that what is true of the pea is true of other seeds, namely, that the root is the first member to make its appearance. We might indeed have guessed that this would prove to be the case; for it is by the root that the plant fixes itself in the soil, and it is an essential condition for the vast majority of flowering plants that they should be firmly "rooted" in the soil.

The next stage in germination may be seen in pea seedlings, the radicles of which are about an inch in length. In such seedlings, a curiously looped structure makes its appearance. One end of this structure is continuous with the radicle, the other end is still in the seed. Gradually the loop lengthens and ultimately its free end becomes visible. This free end, when looked at through a pocket lens, presents a somewhat plume-like appearance, the parts corresponding to the feathers being small, green, flat structures, the smallest of which are very minute and wrapped round the tip of the looped body. By comparing this with later stages, it is evident that the looped body, which soon straightens out and points upward, is the stem, and the small green structures, which arise as outgrowths from it, are the leaves. This seedling stem, with its minute leaves, is called the plumule. Following it downward towards the radicle we find that, at a certain point, there are attached to it, on either side, two stalks. If we remove the seed-coat from the seed of a germinated pea, we discover that the part of the seed which was enclosed by it separates into two halves, and that each part is connected with one of these stalks. The two, massive, more or less hemispherical bodies, which are connected by stalks with the stem of the seedling, are called the cotyledons. Presently we shall have to find out what they are and why they have this peculiar shape.

The different parts of the seedling have received different names. Nor is this unnecessary, for it facilitates comparison between different seedlings and enables us to appreciate the fact that, despite their great differences in shape and size, large numbers of seeds and seedlings are built on the same fundamental lines. That this is so may be seen by comparing the seeds and seedlings of the pea, radish, and sycamore, illustrated in Figs. 2, 3, and 4.

Each of these seedlings consists of a shoot and a root (radicle). The shoot is made up of a leaf-bearing axis, which terminates in a bud (the plumule). That part of the shoot which lies above the insertion of the cotyledons is called the epicotyl, and that part which lies below the cotyledons and merges insensibly into the root is called the hypocotyl. In some plants, *e.g.* pea, the epicotyl makes up nearly all of the mature shoot-system, the hypocotyl remaining short; in others, *e.g.* radish, the epicotyl grows but little. Throughout its first year, it remains very short, and produces a rosette of leaves aris-

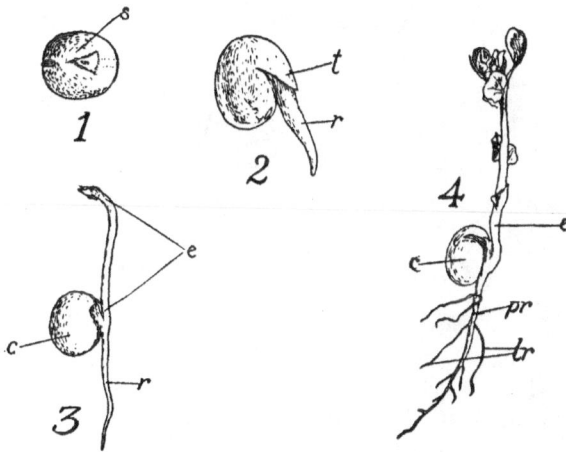

FIG. 2.—GARDEN PEA (PISUM SATIVUM). SEED AND SEEDLINGS.

1. Seed. 2, 3, 4. Seedlings in three successive stages of germination. *s.* Seed; *t*, testa ; *c*, cotyledon ; *r*, radicle ; *e*, epicotyl ; *pr*, primary root ; *lr*, lateral roots.

ing from about the ground level. The hypocotyl, on the contrary, increases considerably in length, and, with the root, forms the edible part of the radish.

By comparing such very different seedlings as those of the pea, mustard or turnip, the castor-oil plant, etc., we reach the conclusion that similar parts are present in each, though they are developed to different extents in the different plants.

We have now to investigate the nature of the two large, lobed bodies which, as we have seen (p. 14), are attached each by a stalk to the stem of the seedling. By re-

moving the coat of a soaked, ungerminated pea or
bean seed we find that, on pressing gently on the
seed along the edge opposite the radicle, the white,
fleshy mass shows signs of separating into two halves.
If we insert the point of a pen-knife in the split we

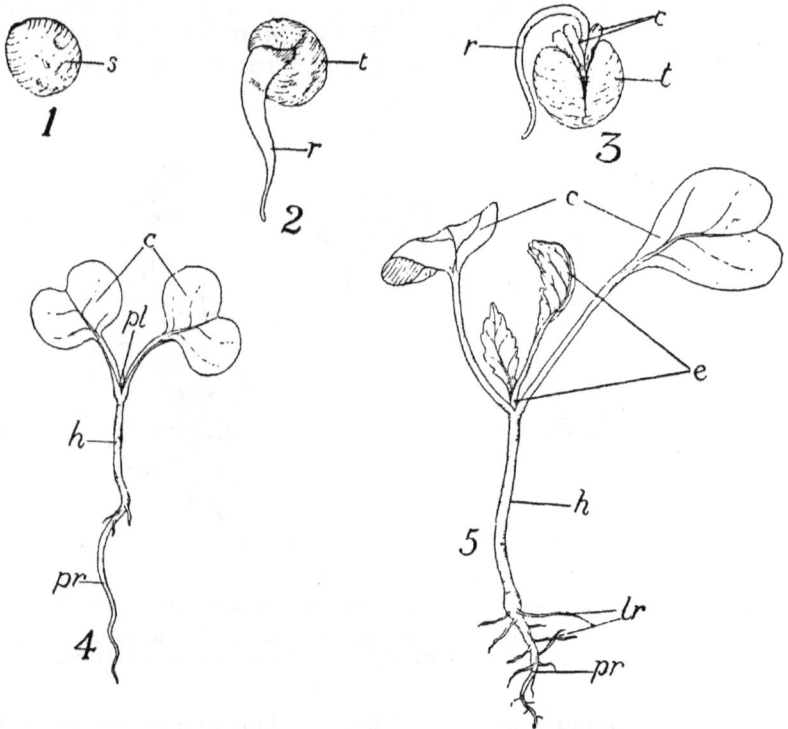

FIG. 3.—RADISH (RAPHANUS RAPHANISTRUM). SEED AND SEEDLINGS.

1. Seed. 2, 3, 4, 5. Seedlings in four successive stages of germination. *s*, seed;
t, testa; *r*, radicle; *c*, cotyledon; *pl*, plumule; *h*, hypocotyl; *e*, epicotyl; *pr*, primary
root; *lr*, lateral roots.

can, without tearing anything, force the two halves
apart and see, by the aid of a lens, that the epicotyl, though
small, is already formed, and lies pressed close to the
inner surface of one of the two fleshy bodies, which
we recognise as the cotyledons. Hence the seed of

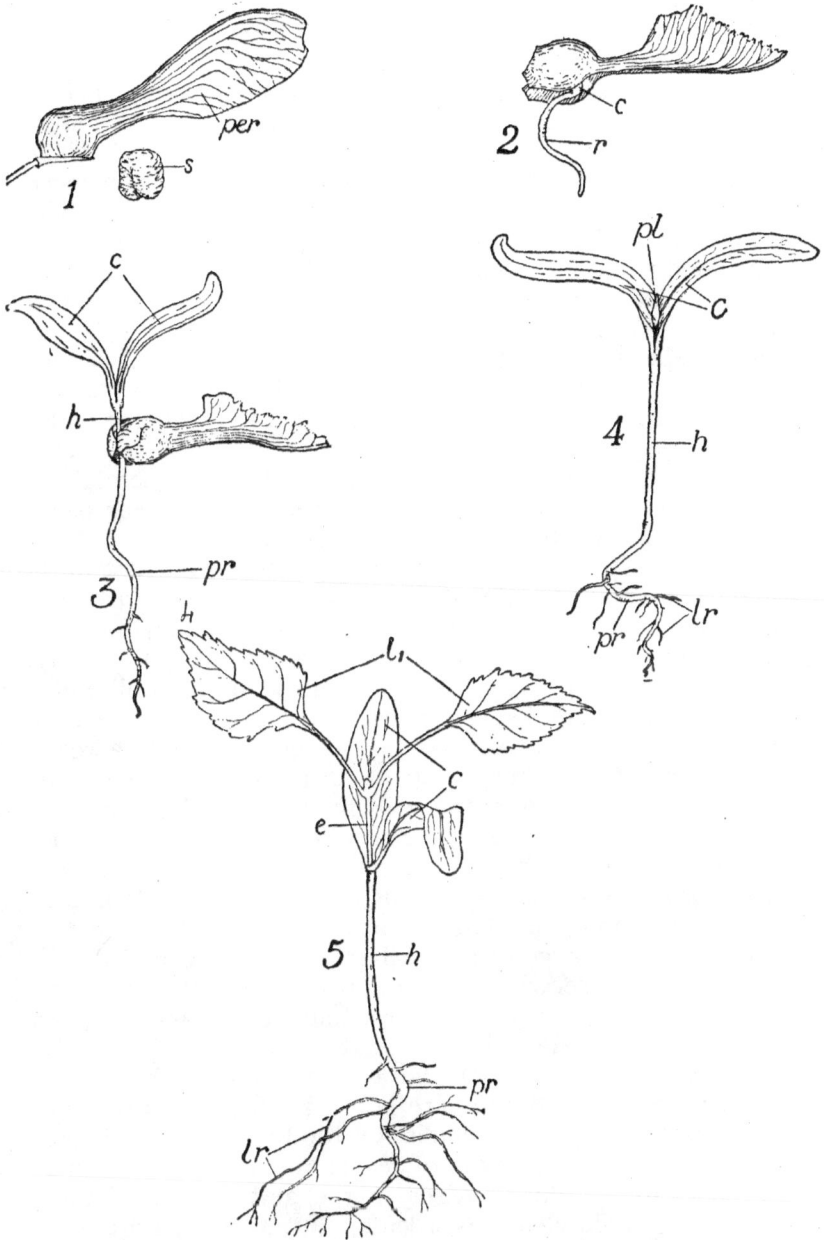

FIG 4.

SYCAMORE (ACER PSEUDO-PLATANUS). FRUIT, SEED AND SEEDLINGS.

1. Half-fruit (mericarp): seed removed. 2, 3, 4, 5. Seedlings in four successive stages of germination. *per*, perianth; *l₁*, first foliage leaves; other lettering as in Fig 3.

the pea consists already of a miniature plant with epicotyl, hypocotyl, radicle, and cotyledons attached to the stem at the junction of epicotyl and hypocotyl. The whole embryo, as the plant in the seed-stage is called, is enclosed in a seed-coat, which, except for a minute hole (micropyle) at the place where the root will emerge, forms a continuous envelope about it. The structure of the pea or bean seed is now clear, except in one particular. We recognise in the embryo all the parts present in the young plant with the exception that nothing comparable with the cotyledons occurs in the latter. We must therefore attempt to discover what the cotyledons are, and what are their particular functions.

First, we will see whether we can find out something about the cotyledons by examining the young pea or bean-plants which we have raised from seed.

Dig up a plant which is from six inches to a foot high, and, observing that the remains of the seed are still attached to the stem, remove the seed-coat and note the shrivelled cotyledons. Sometimes we may see that a bud or even a small branch springs from the angle (axil) which the stalk of a cotyledon makes with the stem. If we do not find such axillary buds or branches, we have a ready means of causing them to grow large enough to be seen; namely, by preventing the growth of the main stem.

10. Thus, having selected a seedling bean-plant about three inches in height, growing in a pot or in the open, cut away or pinch off the stem at the ground level. After some weeks, we note, on digging up the plant, that the two shoots which have formed arise, each in the axil of the stalk of a cotyledon. Now take a branch of a tree and examine it to see how its lateral buds and branches arise. They will be found, in the vast majority of plants, to occur only on the stem just where a leaf is borne; in other words, lateral buds are borne in the leaf-axils. We may argue thus, lateral buds and the branches which they produce arise in the axils of leaves: a lateral bud, which develops to a branch, arises in the axil of a cotyledon: therefore the cotyledon is a leaf. But such an argument, to be convincing, requires to be supported by other evidence. Let us therefore see if further evidence is to be

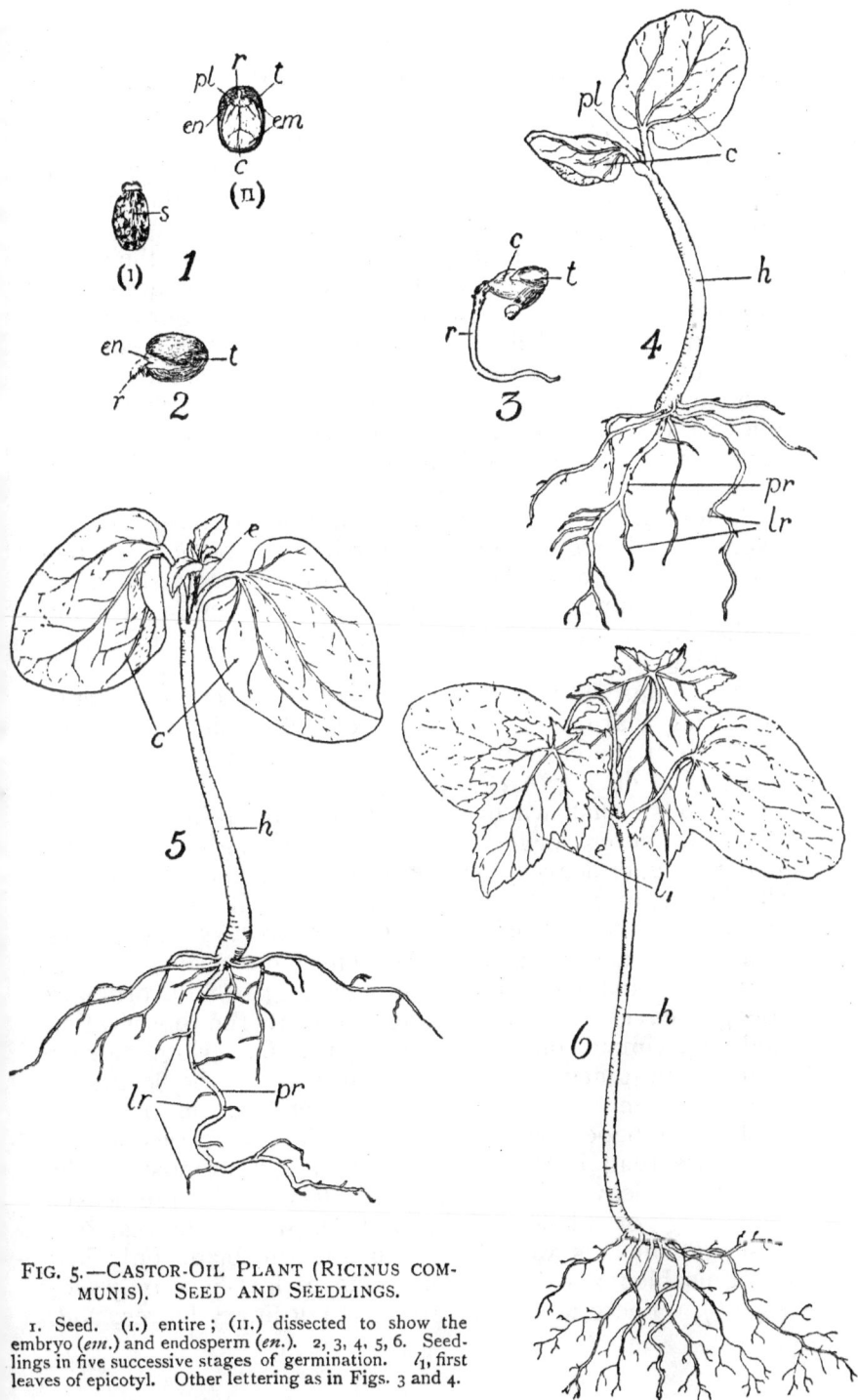

FIG. 5.—CASTOR-OIL PLANT (RICINUS COM-
MUNIS). SEED AND SEEDLINGS.

1. Seed. (I.) entire; (II.) dissected to show the
embryo (*em.*) and endosperm (*en.*). 2, 3, 4, 5, 6. Seedlings in five successive stages of germination. *l₁*, first
leaves of epicotyl. Other lettering as in Figs. 3 and 4.

found. The kidney bean (Phaseolus vulgaris) is very like the broad bean (Vicia faba). They both belong to the same family, and are likely, therefore, to be similar in their essential characters.

11. Sow three or four kidney beans and study their germination. It will be found that they make no concealment of the foliar nature of their cotyledons; for, as germination proceeds, the cotyledons are drawn out of the seed-coat, borne upward above the ground, turn green, spread out flat, and show themselves to be leaves. Cotyledons which rise above the surface of the ground are said to be epigean; those, *e.g.* of the broad bean, which remain below the ground, are termed hypogean.

As an exercise in observation—and observation requires exercise and frequent exercise for its development—determine the nature of the cotyledons of the castor-oil plant (Ricinus communis, Fig. 5), firstly, by dissecting carefully soaked seeds after removing their coats, and, secondly, by observing germination-stages.

It remains to find an answer to the question :—why, admitting that it is a leaf, is the cotyledon apt to depart so much from the conventional form of leaves?

12. If, during early spring, we look at a lilac or privet bush just when the buds are breaking, we find that the outer leaves of the bud, though green and like the inner leaves, remain small, and may fall off as the bud grows to form a branch. But if we cut away the top of the bud with its group of very young leaves just after it has opened, the outermost leaves, which normally remain small and fall away, grow into ordinary foliage leaves (museum). Now repeat the observation and experiment on the buds of the flowering currant (Ribes sanguineum). By dissecting away the leaves of the just-opening bud and laying them out in order on a sheet of paper, we see that they form a series which, when we read it backwards, from the innermost to the outermost leaves, comprises fully-formed foliage leaves consisting of blade, stalk, and leaf-base; leaves, the base of which is well developed and the blade only just recognisable; and leaves consisting of leaf-base only (see Fig. 6). These last never grow into ordinary leaves, and, having served the purpose of protecting the bud during winter, are cast off on the opening of the buds in spring (museum). From these examples we learn that a plant-member, such as a leaf, may be constrained, according to the need of the plant, to change its function. We learn

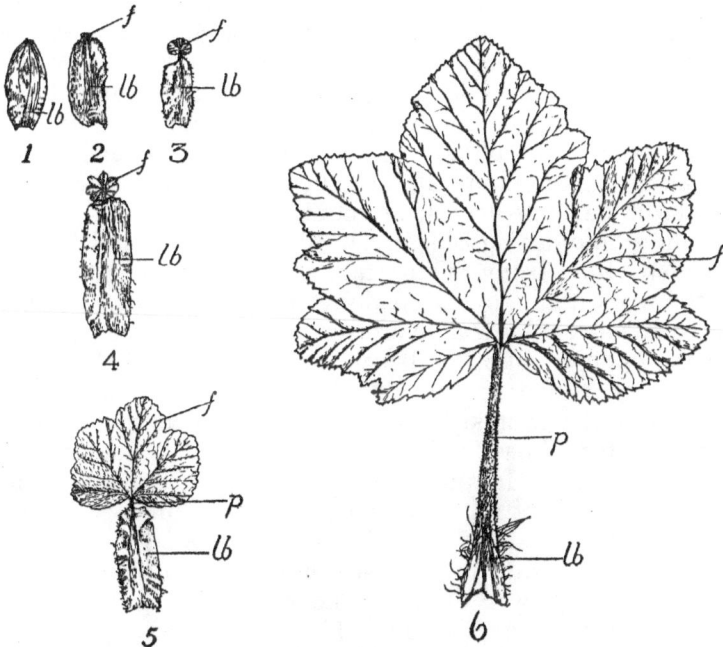

FIG. 6.—FLOWERING CURRANT (RIBES SANGUINEUM).

1, 2, 3, 4, 5. Series of leaves from opening bud, showing transitions from scale leaf (1) to foliage leaf (6). *lb*, leaf-base ; *p*, leaf-stalk ; *f*, blade.

further that when the function changes, the form may also change.

Many important facts follow from this conclusion. For example, it is evident that the particular form assumed by a leaf, or stem, or root, has some relation with the kind of work it has to do. Hence, just as we infer when we see a sharp-edged tool, that it is used for cutting, so, when we have had some general experience of the kinds of work to which the different parts of a plant's machinery are put, we are able to infer from a departure from the normal, usual form displayed by a member of the plant-body that that member is charged with a new, unusual duty. This power of adjusting means to ends is possessed by plants and animals to an extraordinary extent, and is spoken of as adaptability; the modification being called an adaptation. The power of adaptation by which an organism may modify the function of an organ and effect changes of structure and of shape serving to fit the organ for its new work, should be studied by the student at suitable seasons of the year. For example, he should determine the morphological nature of the tendrils of a pea plant, asking himself, are these structures stems, or leaves, or parts of leaves, or roots? When he has solved this morphological problem, let him ask himself whether this departure from the normal structure fits the organ for the work it has to do. Among the innumerable subjects for such morphological exercises we may mention potato tubers, opening buds of beech, onion or hyacinth bulbs, strawberry runners, and double flowers (stocks, roses, etc.); but the best subjects are those which the student discovers for himself. The method he must use is that of comparison. He must compare the thing with itself at different stages of its development, *i.e.* he must study its developmental history or embryology, and he must compare the thing with its nearest allies, *e.g.* a double flower with a single flower of the same species—a garden rose with a wild rose, and so on. This is the method of comparative anatomy, which may be pursued further and made to include microscopic as well as a naked-eye investigation.

Now to apply our conclusions to cotyledons. In some plants, the foliar nature of the cotyledons is obvious : *e.g.* in plants of the cabbage tribe—turnip, cress, cabbage, etc.—and in the castor-oil plant, etc. ; in others, though not apparent at first, it becomes so during germination, *e.g.* kidney bean ; in yet others, *e.g.* the pea, bean and horse-chestnut, it is by no means apparent. Keeping in mind the cases of the buds of the lilac or privet and of the flowering currant, we surmise that the reason why some cotyledonary leaves have retained their foliar character whilst others have, in large measure, lost it, is that the former have retained the functions of ordinary leaves, and that the latter have exchanged these functions more or less completely for others. Change of form is an outward and visible sign of change of function. What then is the new function which the thick, unleaf-like cotyledons of the pea have assumed? When the castor-oil seed was dissected, the student must have been struck by the fact that the embryo did not take up the whole of the space within the seed-coat. He saw that, covering the thin, delicate seed-leaves (cotyledons), which lie pressed together in a plane median to, and parallel with the flatter surfaces of the seed, there is a mass of soft, white tissue. This tissue, called endosperm, from which the castor oil of commerce is extracted, has no counterpart in the ripe pea or bean seed. On the other hand, whilst the cotyledons of the pea and bean are thick and unleaf-like, those of the castor-oil seed are thin, and so leaf-like as to show even their "veins." Let us state the facts thus :—Castor-oil seed; cotyledons thin and leaf-like; a mass of tissue (endosperm) external to the embryo making up the larger part of the seed. Bean seed; cotyledons fleshy, no corresponding endospermous tissue external to the embryo; the seed-leaves making up the larger part of the seed. If we could discover the use of the endosperm to the seedling, we could make a good guess why the cotyledons of the bean are fleshy. Conversely, if we could find out what purpose is served by the fleshy cotyledons, we should know probably why the castor-oil seed contains endosperm. Again, as always, when confronted with problems in plant-physiology we have recourse to our scientific method of guess and experiment. Now it cannot have escaped our notice when we were germinating various kinds of seeds that some are small and produce small seedlings, and some are large and produce large seedlings. If we examine seeds of the latter kind, we see that they have invariably either large and fleshy cotyledons or much endosperm, and if we dissect seeds of the former kind we find that the cotyledons are thin and that, if any endosperm is present, the amount is but small.

As a study of the museum specimens shows, the size of a seed has a relation with the size of its cotyledons

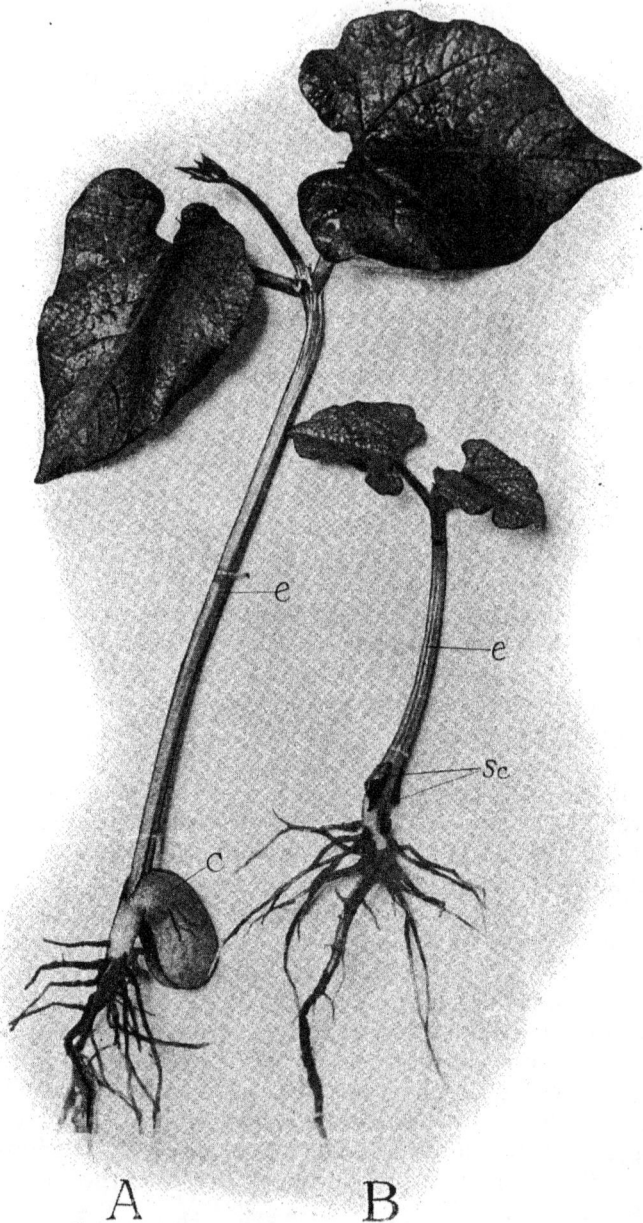

FIG. 7.—KIDNEY BEAN (PHASEOLUS VULGARIS).
From a Photograph.

A. Seedling germinated under normal conditions. B. Seedling from seed sown at the same time as A, but from which the cotyledons were removed at germination. *c*, cotyledon ; *e*, epicotyl ; *Sc*, scars left after removal of cotyledons.

or endosperm, and not with the size of the mature plant which it produces. Moreover, a large seed produces a large seedling, and a small seed a small seedling. We can scarcely doubt that the big seedling is due in some way to the big cotyledons or to the large amount of endosperm, and, thinking of the effect which proper feeding has on the size of young animals, the idea dawns upon us that perhaps the large cotyledons and the endosperm contain food supplies on which the seedling is enabled to feed during germination. This hypothesis we can put to the test. If it is true, then, as the seedling grows, the cotyledons should be found to shrivel and the endosperm to disappear. We have already seen the shrivelled cotyledons of the young bean plant (p. 18), and so have some confirmation of the correctness of our hypothesis.

13. To obtain complete proof we may proceed to weigh the shrivelled cotyledons of a bean which is about a foot high. After drying them in a drying oven, we compare their weight with that of the cotyledons of ungerminated bean seeds, similarly dried. We find that the dry weight of the shrivelled cotyledons is considerably less than that of the fresh cotyledons of ungerminated seeds. Another and more striking method : as soon as the epicotyl appears in each of six kidney bean seedlings sown in germinators, remove, by means of a sharp knife, the cotyledons from three of them, and plant all six seedlings in a pot with garden soil. Determine the rate of growth of the two sets (Fig. 7). Record : draw and preserve one plant of either set for the museum. (For this purpose the plants may be dried between blotting-paper and mounted on cards.) A similar experiment may be made with maize or wheat seedlings, in which case it is the endosperms which must be removed (cf. Fig. 8). The result in either case is the same. The seedlings formed from the seeds deprived of their fleshy cotyledons or of their endosperm, though they grow, increase in size much more slowly than those from intact seeds. Whence we conclude that endosperm and fleshy cotyledons serve as reservoirs of food-materials on which seedlings draw for their nutrition. Presently (Chapter III.) we shall have to enquire into the nature of these food-materials and how the seedling obtains them.

It is interesting and typical of the ways of living things that the same end, in this case that of endowing the seedling with a capital of food wherewith to start in life, may be secured by different ways: the one, by converting the

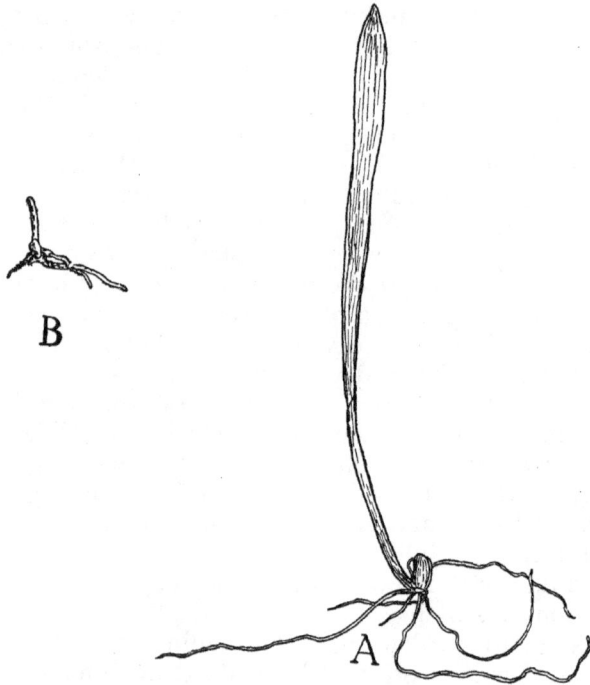

FIG. 8.—WHEAT (TRITICUM SATIVUM).

Showing the relative sizes of seedlings grown with and without endosperm.

A. Seedling, germinated on moist blotting-paper. B. Seedling of the same age as A, germinated under similar conditions, except that the endosperm was removed as soon as germination began.

seed-leaves into storehouses, the other, by the provision of a separate, nutritive tissue (endosperm) external to the embryo. In point of fact, the difference between seeds with, and seeds without endosperm only amounts to this, that the embryo of the endospermous seed waits until it begins to grow before taking up the provision of food-material made for it by the parent plant; whereas the embryo of the seed without endosperm takes up the food-substances and stores them in its cotyledons before it ripens and comes to rest. If the parent makes a liberal provision, the cotyledons become thick and fleshy; if the provision is niggardly, the cotyledons remain thin and the seed is small. With a knowledge of the meaning of big cotyledons and large masses of endosperm, we obtain an understanding of the significance of the great variability of seed-production shown by different plants. The parent plant provides the material out of which the tissues of the cotyledons or endosperm and embryo are formed. The parent may produce either many small seeds or fewer large seeds : compare, for example, the poppy and pea. The one or the other habit runs in plant-families, and it is interesting to note that, of two of the most successful plant-families, the members of one, Leguminosae (peas, beans, clover, vetches, etc.), produce large and relatively few seeds; those of the other, Compositae (daisy, dandelion, etc.), produce small and relatively many seeds. We can also understand that the work of seed-production entails a certain amount of exhaustion to the parent plant, and that, therefore, a copious crop of seeds and fruits one year may use up so much of the food-material at the disposal of the parent that but few, or even no seeds may be produced the following year, e.g. beech, etc. So much is this the case that there is a great group of plants which die after once flowering. To this group belong our annual garden plants and weeds which live for one season or less, produce seed and die. If they are prevented from flowering, their lives may be prolonged, as any one with a garden or even a space for pot-plants may prove for himself. Not all once-flowering plants, however, are annuals, various tropical or sub-tropical shrubs and even trees grow for many years and ultimately flower, set seed, and die (e.g. species of aloë, palms). Other plants are biennials, that is, flower in their second year (turnip, carrot, foxglove); whilst others again are perennials, and may live for centuries and flower again and again.

Having accomplished our objects, of learning something of the general nature of plants and of beginning our studies on nutrition, we will conclude this chapter with the remark that the pea or bean type of seedling, with its axis terminating above in a plumule and below in a radicle and bearing two cotyledons laterally, is not the only type of seedling to be met with among flowering plants. If we call this type the Dicotylous type, we may say that there is another, very varied type, viz. the

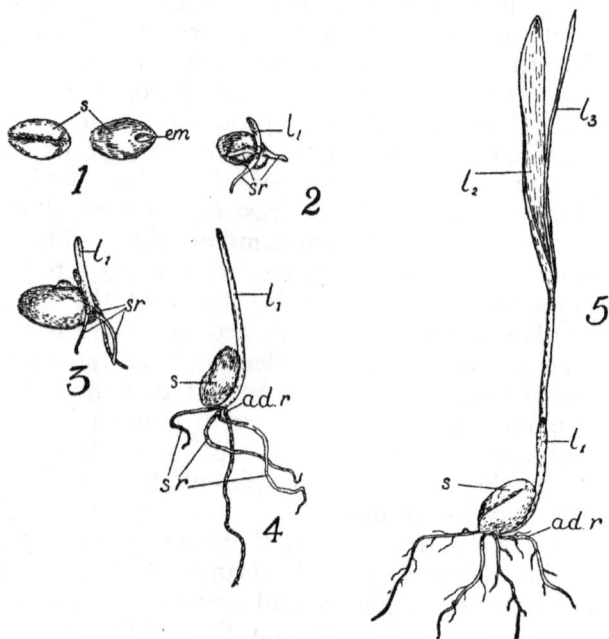

FIG. 9.—WHEAT (TRITICUM SATIVUM). FRUIT AND SEEDLINGS.

1. Grain of wheat (*i.e.* a fruit enclosing one seed). 2, 3, 4, 5. Seedlings in four successive stages of germination. *s*, fruit; *em*, embryo seen through thin adherent coats of fruit and seed; *sr*, 'seminal' roots; *adr*, adventitious roots of permanent root system; *l₁*, *l₂*, *l₃*, 1st, 2nd and 3rd leaves.

Monocotylous type. Seeds of the latter, *e.g.* onion, wheat, oats, barley, maize, and date, should be examined, germinated, their seedling stages studied, and preserved in the museum (Fig. 9). Some help will be required in making out the parts of such seeds as these, but this may be obtained from a text-book of general botany (Bibliography, 3, 5).

CHAPTER III.

THE nature and chemical properties of the food-substances contained in the cotyledons and endosperm of seeds.

WHEN we reflect on the results of our observations and experiments on fleshy cotyledons and on endosperm, we are bound to be struck by the fact that those seeds which contain large quantities of food-material serving for the nutrition of the seedlings are also the seeds which are used most largely by man and animals for food. When cereal crops are ripe, that is, when the plants have transferred stores of food-materials to the endosperm of the seed, man intervenes and, gathering in the crop, makes flour from the grains. Flocks of birds anticipate man, and by their depredations bring serious loss to the farmer. Teeming populations in the East support life solely on rice, the seed-like fruit of a grass (Oryza sativa). The oilcake on which cattle are fattened is derived from the remains of the reserve-materials of the seeds of rape, cotton, etc. Thus the conclusion impresses itself upon us that the food-materials of the seedling serve also as food for man and animals. That this is so, will add interest to our present enquiry into the nature of these food-materials and the way in which they are used by the seedling. Since a thorough examination into the chemistry of the food-materials, or, as we may call them, the reserve-materials of the seed, involves both a knowledge of chemistry and also the occasional use of the microscope, and since some students may lack the necessary chemical knowledge or be unskilled in the use of the microscope, we will indicate by means of an asterisk (*) the experiments which may be omitted by beginners.

14. Soak one or two wheat or barley grains in water : when soft, cut them in halves, and smear some of the white endosperm on a saucer or porcelain slab : add a few drops of a solution of iodine dissolved in water with potassium iodide (Appendix A); note that a blue colour is produced. Obtain some finely powdered, pure starch from a chemist; repeat the potassium-iodide iodine test (1) by adding the solution direct to the powder; (2) by first boiling a little of the starch with water in a test tube or beaker, then diluting the boiled liquid with several times its volume of water, and, after cooling, adding the iodine solution. Heat the

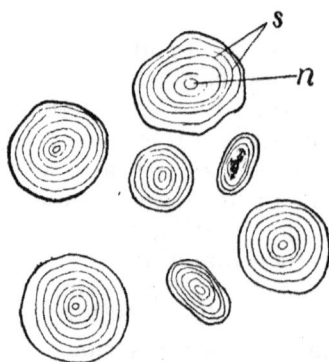

FIG. 10.—STARCH GRAINS FROM THE ENDOSPERM OF BARLEY.

n, hilum of grains ; *s*, lines of stratification due to differences in constitution of successive layers.

blue liquid, note that the colour disappears : cool it under the tap, note that the colour reappears. We thus determine that the blue colour obtained in the previous experiment is due to starch. We speak of this reaction as a test for starch, or as the *iodine-reaction* for starch.

15.* Scrape the cut surface of a soaked bean seed or barley grain with a clean knife, transfer a trace of the scrapings to a small drop of water on a glass slide, put on a cover glass and examine the preparation at first with a low and then with a high power of the microscope. Draw several starch grains, indicating the characteristic concentric markings (Fig. 10).

16.* Cut sections through the endosperm of wheat, the

cotyledons of the bean and the tuber of the potato.   Note
the positions of the starch grains, groups of which lie in
the compartments of which these tissues are composed.
Draw.   Run in under the cover glass of one of these
microscope-preparations a drop of potassium-iodide iodine
solution.   Note the blue or blue-black colour of the grains.
Crush soaked wheat grains in a mortar with a little water,
transfer some of the pasty mass to a test-tube: shake
well and filter through a filter funnel lined with filter

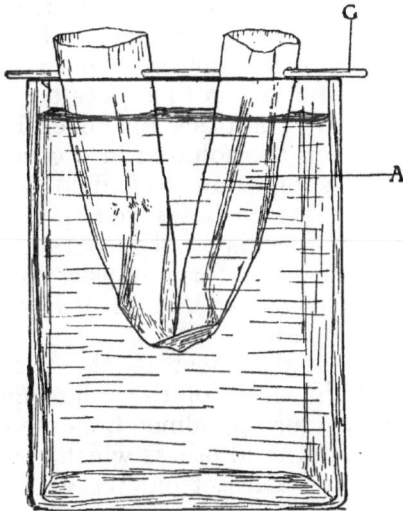

FIG. 11.—DIALYSER.
A, parchment tube ; G, glass rod.

paper.   Determine that the filtrate, *i.e.* the liquid which
passes through the filter paper, gives no blue colour
with the iodine solution; in other words, that starch is
insoluble in water.

   17.   Boil some commercial starch with water : note that
it passes into an opalescent, thin, jelly-like condition, but
does not dissolve.   Dilute some of this starch-containing
liquid with water and pour it into a parchment shell or
tube of parchment paper (Appendix B), bent in the
form of a **U**, and supported in a beaker of water by
means of a glass rod passing through the upper part of the
limbs of the **U** (Fig. 11); leave it for twenty-four hours.   If

starch is a diffusible substance it will pass through the parchment paper into the water in the beaker. Prove by the iodine test that starch is not diffusible.

18. Add a little starch—as much as will go on to the end of a knife blade—to about a half-pint of water in a beaker, stir well, and boil: add by means of a glass rod two or three drops of a mineral acid, *e.g.* hydrochloric acid, boil, and at hourly or convenient intervals test the liquid for starch thus :—by means of a glass rod, distribute drops of the liquid on a porcelain saucer, dip another clean glass rod into the iodine solution, and mix drops of the latter with those on the saucer. Observe that the starch reaction becomes fainter as time goes on, and finally fails. Boiled with small quantities of mineral acid, starch is decomposed. Keep the solution in a clean bottle (for use in Exp. 25). It is important to find out to what substance or substances the starch has given rise. We may get a hint with respect to this in the course of the following experiments which we perform in order to find out if starch is generally present in seeds, and if it is widely distributed in other parts of plants.

19. Apply the iodine test to the cut and moistened surfaces or to thin slices of the following seeds : Brazil nut, date, maize, turnip, almond, horse-chestnut, cress, rape, castor-oil plant, and others which can be spared from the seed-collection. Dissect out the embryos of just ger-minated barley or wheat grains, put them for about half an hour into a solution consisting of equal parts of saturated chloral hydrate and potassium-iodide iodine (Appendix A). Note that the embryos give a beautiful blue starch-reaction.

20. Cut thin slices of unripe and ripe apples or pears and compare them as to starch content. Test thin slices of the stems and roots of various flowering plants for starch. Find out whether starch is present in green leaves, *e.g.* of clover, lime, lily, American water-weed (Elodea canadensis), snowdrop, iris, and other plants. For this purpose, proceed as follows : throw the leaves, which should be thin, into boiling water for a minute or two. Take them out and put them into a wide-mouthed bottle containing methylated spirits : stopper or cork the bottle, and expose it to sunlight: after a day or two, remove the colourless and brittle leaves carefully from the alcohol, wash them in water, and pour over them chloral hydrate, iodine solution as used in the previous experiment. Record which leaves contain starch, and which do not. Test similarly pieces of fern fronds, moss leaves, the green threads of any algæ which may be found growing in ditches, and pieces of mushroom or of toadstools.

When ripe and unripe fruits, apple, etc., were tested for starch, it was noticed that the unripe fruit contained much starch and the ripe fruit far less or even none. On the other hand, the ripe fruit is sweet, and the unripe fruit is not. It therefore appears likely that the sweetness of the ripe fruit is due to the conversion of starch into sugar. This is rendered the more probable in that both starch and sugar consist of the same three elements, viz. Carbon (C), Oxygen (O), and Hydrogen (H), and have this further in common, that the proportion of hydrogen to oxygen in their respective molecules is as 2 : 1 (just as it is in water, the chemical formula of which is H_2O). Bodies having this constitution are classed in the chemical group of the carbohydrates. The differences and similarities of the constitution of the molecules of starch and of a sugar may be seen if we write down their respective formulæ :

Starch, $(C_6H_{10}O_5)_x$.   Sugar (grape sugar), $C_6H_{12}O_6$.

The x outside the bracket in the starch formula means that a starch molecule has not 6 carbon, 10 hydrogen, and 5 oxygen atoms, but some multiple of these numbers, e.g. x may equal 100 or more (its exact value is not certain), and assuming that it is 100, the composition of the starch molecule is $C_{600}H_{1000}O_{500}$. If we neglect this complication of the bigness of the starch molecule, and write it $C_6H_{10}O_5$, then it is evident that it only differs from the grape sugar (glucose) molecule by H_2O; in other words, if we could add a molecule of water to one of grape sugar we could represent the conversion of starch into grape sugar thus: $C_6H_{10}O_5 + H_2O = C_6H_{12}O_6$. When water is induced to combine with such a body as starch, the change undergone by that body is described as one of hydrolysis. These considerations with respect to the chemical constitutions of starch and sugar show that there is no inherent improbability in the view that starch, when it disappears from the ripening fruit, may do so because it has been converted into sugar.

Inasmuch as sugars are widely distributed in plants, and evidently play a part in plant- as well as in animal-nutrition, we must consider their properties. We determine that they are soluble, using grape sugar (glucose) and cane sugar (sucrose) for the experiments. We also require a ready test for the identification of this important class of substances. The test we employ is based on the fact that some sugars have the power of taking oxygen from certain substances, and hence of reducing these substances to a less oxidised condition. For example, such reducing sugars, under suitable conditions, take oxygen from cupric oxide (CuO), reducing it to cuprous oxide (Cu_2O), thus :

$$4CuO = 2Cu_2O + O_2,$$

and inasmuch as cuprous oxide is an insoluble, yellow-red substance, we are able to see if it is produced when a solution of sugar is added to one containing cupric oxide. Thus, we have a basis for a test for reducing sugars. This test we apply as follows :

21. Pour the liquid to be tested into a test tube or beaker : add an excess of potash and then a drop or two of a solution of copper sulphate. The precipitate of copper hydroxide $Cu(OH)_2$, which is formed by the interaction of copper sulphate and potash is dissolved by the excess of potash. Boil : if a reducing sugar is present, a yellow-red precipitate of cuprous oxide is produced. Dissolve grape sugar (which is readily obtainable in commerce) in water and apply the above test. Repeat, using cane sugar : observe that grape sugar is a reducing and cane sugar a non-reducing sugar.

22. Pour another portion of the solution of cane sugar into a beaker and add a drop of hydrochloric acid ; boil for some time, and apply the potash and copper sulphate test ; observe that a well-marked precipitate is formed. It is evident from the experiment that cane sugar, when boiled with a trace of mineral acid, yields a reducing

sugar. The process whereby a non-reducing is converted into a reducing sugar is called inversion. In
the case of cane sugar, the chemical change may be
expressed thus :

$$C_{12}H_{22}O_{11}+H_2O=2C_6H_{12}O_6.$$

The change is one of hydrolysis, that is, one in
which water is caused to combine with the original
body, in this case cane sugar. Instead of using potash
and copper sulphate separately in testing for reducing
sugars, a ready-made solution called Fehling's solution,
which produces the same result (Appendix A), may be
employed.

23. Test by means of potash and copper sulphate or
Fehling's solution the following plant tissues or extracts
therefrom : ripe and unripe apple, beet, carrot, onion.

24. Crush with water in a mortar grains of barley
which have been soaked in water for one or two hours :
filter the extract, label it Extract A. Now crush grains of
barley which, germinated in a germinator, are showing
leaves an inch or two long; prepare a watery extract :
label it Extract B. Test samples of A and B for sugar.
If neither gives a distinct reaction, concentrate A and B
each to a small bulk by heating them in porcelain dishes
over a bunsen flame or water bath and then repeat the sugar
test. It will be found that the extract of the soaked,
ungerminated grains contains no sugar (or at most a trace),
whereas the extract of the germinated grains contains a
considerable amount. Prove that the sugar is contained
in the endosperm. It looks certain from these results, and
also from those obtained by testing ripe and unripe fruits
(apple, etc.), that the starch contained in the ungerminated
barley and in the unripe fruit becomes converted during
germination and during ripening into sugar. We have a
means at hand of demonstrating that starch *does* give rise
very readily to sugar.

25. Take the liquid obtained by boiling starch with
water and a mineral acid (Exp. 18). Test a sample in a
test tube for sugar. If no marked reaction is obtained,
concentrate the remainder of the liquid to a small bulk and
repeat the sugar test. Learning thus that boiling with a

trace of mineral acid suffices to hydrolyse starch to sugar, we shall be the more ready to believe that the sugar which appears in ripening fruit or germinating barley is produced by the hydrolysis of starch.   Presently we shall have to determine what agent in the tissues of the plant is responsible, like the mineral acid of Exp. 18, for this conversion of starch to sugar, and of what significance to the seedling is this conversion.   When we do this we shall require to know the various properties which distinguish sugars from starch.   One such property, that of solubility in water, is well known.

26.   Another, that of diffusibility through a parchment membrane, is no less important, and must be demonstrated by the use of the dialyser (Exp. 17).   Some hours after the sugar solution has been placed in the parchment tube, the liquid contained in the vessel in which the tube hangs is tested for sugar either directly or after concentration to a small bulk.   The result of the experiment proves that a solution of sugar separated from water by a parchment membrane passes across the membrane into the water.   This passage is called osmosis, and substances possessing this property are called osmotic substances (see Chapter vii.). Let us now refer to our list of seeds tested for starch (Exp. 19), and note those from which starch was found to be absent.   They should include the following :—rape, castor-oil, almond, date, Brazil nut.   But these seeds all have either endosperm or fleshy cotyledons, and it is therefore more likely that they contain some form of reserve food-material other than starch, than that they contain no reserve food-material at all.   A scrutiny of the list shows us that certain of these seeds contain fat or oil; indeed, as we know, they owe their use in commerce to this fact. We will select one kind of seed from the above list for examination, *e.g.* the castor-oil seed (Ricinus communis).

27.   Pound a seed in a mortar with a few drops of ether or benzol (Appendix A) (this must not be done near a flame, for ether and benzol are highly inflammable; it is safest to do it out of doors).   Pour the ether or benzol extract on a sheet of paper.   Note that, as the liquid evaporates, it leaves behind a grease spot.   To the grease spot or to the remains of the crushed seeds, add a drop or

two of osmic acid (Appendix A). Note that the osmic
acid produces a black or brown colouration. Repeat the
test on a drop of salad or linseed oil in a white saucer.

28.* By means of a dry razor or knife cut thin
sections of the endosperm of the castor-oil seed, mount
them on a slide, examine them microscopically and run in
osmic acid : note the brown-black masses of oil.

29. Now examine, by means of the above tests, the
remains of the endosperm of seedlings of the castor-oil
plant in different stages of germination : note that, as
germination proceeds, the oil disappears from the endosperm. From analogy with the occurrence of starch in
other seeds and with its behaviour during germination, we
conclude that oil and fat are the forms in which some seeds
store their reserves of food. Like the carbohydrates, fats
and oils contain the three elements, carbon, oxygen, and
hydrogen ; but whereas in the carbohydrate molecule there
are twice as many atoms of hydrogen as there are atoms of
oxygen, in fats the number of hydrogen atoms per molecule is more than twice the number of oxygen atoms.
For a fuller description of the chemical and physical properties of fats, and also of sugars, the student should
consult a text-book of Organic Chemistry (Bibliography,
10, 11).

Among the seeds which were germinated in the
course of experiment (Exp. 19), and tested for starch
(Exp. 26), was the date—the seed of the palm, Phœnix
dactylifera. The seed itself is remarkable. It germinates
very slowly, the endosperm is flinty hard, and the embryo,
placed about the middle of the length of the seed on the
side opposite the groove, is extremely small, and shows,
in the ungerminated condition, no distinction into the
usual parts. Germination-stages of the date seed should
be obtained, and put up in museum jars. We are not now,
however, concerned with the morphological peculiarities
of the seed, but with the nature of the reserve materials
contained in its endosperm.

30.* If a thin section of the endosperm of a soaked date
seed is examined in a drop of water under the microscope,
it will be found that the walls which chamber up the tissue
into a number of compartments are extraordinarily thick.

If now a similar section is made through the soft endosperm of a date seed which has been germinating for some
months it will be observed that the walls, so thick in the
ungerminated seed, have become thin. It might be supposed that this was merely a symptom of decay; but it is
equally open to us to infer that the disappearance of the
substance of the walls is evidence that this substance consists of reserve food-material which serves to nourish the
seedling. This substance does not give the iodine reaction
for starch; but if a thin section of endosperm is first treated
with strong sulphuric acid and then with iodine solution
(Appendix A), a blue colour, like that given by starch, is
produced. This substance of the walls, insoluble in water
and giving a blue colour with sulphuric acid and iodine, is
called reserve-cellulose. It may be found not only in the
seeds of palms, but in others such as lupine and Ornithogalum, seeds of which plants, both ungerminated and
germinated, should be examined.

31.* If the sulphuric acid and iodine test is applied to
sections through the soft parts of plants free from starch, or
even to small pieces of plant tissues, *e.g.* grape or apple, a
blue colour, like that shown by the date endosperm, is
produced, and if sections are treated in this way and
examined microscopically, it is found that the walls
which chamber up the tissue into compartments consist of
cellulose substances. Though the cellulose of the cell-
walls of plant tissues gives the same sulphuric acid and
iodine reaction as the reserve-cellulose of date and lupine
seeds, it is not identical with this latter substance. The
chief difference between ordinary cellulose and reserve-
cellulose consists in this, that the former undergoes
chemical change less readily than the latter. Indeed the
function of the cellulose substances of the walls of mature
tissues is not to serve as reserve food-material; but to act
as a scaffolding to the compartments of these tissues.

Having made the discovery that the seeds of plants contain stores of carbohydrate or fatty material which serve in
some way or other for the nutrition of the seedlings, we
might conclude that our task of investigating the functions
of endosperm and cotyledons was at an end. But if we
remember to apply the principle which we have laid down,

and assumed to be true, that the modes of nutrition of
plants and animals are fundamentally alike, we shall be
struck by the fact that, so far, we have discovered in the
seed no reserve food-substance of a kind similar to that
which is so characteristic of the eggs of animals.   For
example, the hen's egg contains substances which have the
property of setting to a solid mass (coagulating) when they
are heated.   We know from experience that these substances are highly nutritious, and since they disappear
from the yolk during the hatching of the eggs, we may be
fairly certain that they contribute to the formation of the
developing chick.   We must therefore first consider the
nature of these coagulating substances, and then determine
whether similar bodies occur in the seeds of plants.

32. We prepare a solution of white of egg by crack-
ing the shell of a fresh egg and letting the white fall
into a dish, beating it with five to ten times its volume
of water, and filtering.   Heat one part of the solution in
a test-tube : note that a white coagulum is formed.   A
similar precipitate is produced by the addition of alcohol
to the white of egg solution.   By the use of the dialyser
(Exp. 17) we demonstrate that the solution made from
white of egg does not pass across the membrane, i.e.
though soluble, it is not diffusible.   Apply the following
tests to samples of the white of egg solution in test
tubes :

33.* *Xanthoproteic reaction:* add a few drops of
strong nitric acid : a white precipitate is produced, which
becomes yellow on heating : add ammonia cautiously : the
colour of the precipitate changes to orange (Appendix A).
Biuret reaction: add a *trace* of copper sulphate solution,
then caustic soda or potash : a violet colouration is produced.   By the substitution of ammonia for the soda or
potash, a reddish violet colour is obtained (Appendix A).
Millon's reaction: add Millon's reagent (Appendix A);
a white precipitate is formed, which becomes brick-red on
boiling.

Iodine reaction: add iodine, a yellow brown colouration
is produced : the tint being considerably deeper than when
a similar amount of iodine solution is added to pure water.

The group of chemical substances which have the above

properties and give these reactions is called the proteins, and the chief protein contained in white of egg is called albumin (Appendix A). If, as we have already reason to believe, the proteins contained in eggs serve as reserve food-material we shall expect to find them also in the bodies of animals.

34.* That they do occur in the adult body we prove by mincing fresh meat and pounding the fragments in a mortar with water. The extract is then filtered and the filtrate tested as in Exp. 33. We determine also that proteins occur in milk by diluting fresh milk with water and adding *dilute* acetic acid. Filter and test the precipitate for proteins. The curdling of milk is evidently connected with the presence of proteins.

The results of chemical analysis of the proteins show that beside carbon, oxygen, and hydrogen, they contain nitrogen and sulphur, and that some also contain phosphorus.

35.* Demonstrate that proteins contain carbon, nitrogen, and sulphur thus : heat chopped, coagulated white of egg in a porcelain dish : note that it becomes dry and subsequently chars—an indication that it contains carbon. Chop finely some pieces of the white of a hard-boiled egg, dry in a desiccator (or better, use dry, powdered albumin instead, see Appendix A) : grind in a mortar with soda-lime (Appendix A), transfer to a short tube of hard glass closed at one end. Heat in a bunsen flame, if necessary using the blow-pipe. Note the smell of ammonia : hold a piece of red litmus paper over the open end of the tube, observe that it becomes blue. Since ammonia (NH_3) is produced, albumin must contain nitrogen. A piece of lead acetate paper blackens when held in the fumes escaping from the heated tube; the blackening being due to the formation of lead sulphide (Appendix A); hence proteins contain sulphur. We recollect that a silver spoon left long in an egg becomes black (owing to the formation of sulphide of silver), and we recall the smell—of sulphuretted hydrogen—of a bad egg.

To study further the properties of the proteins is beyond the scope of our present work (Bibliography, 10, 11). We can only say that they may be classified according to their properties, particularly according as they are soluble in water (albumins); insoluble in water but soluble in dilute solutions of such salts as sodium chloride, magnesium sulphate, etc. (globulins); soluble only in strong salt-solutions; insoluble in any of these reagents (*e.g.* coagulated albumin). One substance with essentially protein-like properties, but with marked peculiarities, must be mentioned, and its reactions noted.

36.* Take a small quantity of peptone (which is obtainable in commerce, Appendix A), add water. Note that it dissolves. Heat : note that it does not coagulate. Apply the xanthoproteic and Millon's tests for proteins. Test a solution of peptone and another of albumin by means of the biuret test. Compare the violet reaction of the albumin with the rose-pink reaction of the peptone solution.

37.* Demonstrate by means of the dialyser of parchment tube that peptone, unlike proteins, is capable of osmosis. In performing this experiment, which will last several days, owing to the fact that the rate of osmosis of peptone is slow, a *trace* of an antiseptic such as thymol or eucalyptus oil (Appendix A) should be added to a strong solution of peptone. At daily intervals, withdraw some of the liquid from the outer vessel and test a sample of it for peptone by means of the biuret test. If, owing to the diluteness of the solution, no reaction is obtained, concentrate to a small bulk and repeat the test.

We learn from our study of the proteins that they are complex bodies containing the elements Carbon (C), Oxygen (O), Hydrogen (H), Nitrogen (N), Sulphur (S)—some also contain Phosphorus (P)—that they have remarkable properties (of coagulation, etc.), that they may occur in the eggs and also in the mature bodies (*e.g.* the flesh) of animals (Exp. 34), and that there is good reason to believe that they serve for the nutrition of animals. We have in the next place to enquire whether the seed and the mature plant also contain proteins.

38.* Cut thin sections of the tip of a root or stem of any young plant or of the young stamens of a lily or other flower. Mount in water on a slide in the usual way. Examine by means of the microscope. Run in iodine, and note that the contents of the compartments into which these tissues are chambered give the yellow-brown reaction for proteins.

39.* Repeat Exp. 35, using, instead of white of egg or dry albumin, chopped and dried grass, or pea or bean meal.   Note that these vegetable tissues yield evidence that they contain nitrogen and sulphur.   Thus from the results of the last two experiments we infer that proteins are present in plants.   Confirm this as follows :—

40.* Extract pea or bean meal or soaked and pounded wheat grains (1) with water, (2) with a dilute (5 %) solution of magnesium sulphate : filter, and demonstrate by means of the protein tests applied to the filtered extracts that proteins are contained in the seeds of bean or pea and wheat.

41.* Put a little flour in a fine muslin cloth folded to form a bag.   Hold the muslin bag under a tap, and allow water to run on the flour, which should be kneaded or squeezed between the fingers.   Observe that the starch is washed away, leaving a sticky mass of gluten.   Demonstrate that this residue contains proteins (Exp. 33).

Hence we conclude that the reserve-materials contained in the seeds of plants are of like nature to those contained in the eggs of animals, and consist not only of carbohydrates (or fatty substances), but also of proteins.

42.* In order to determine the condition of, and the place in the seed occupied by, the protein reserve materials, we cut, preferably by means of a dry razor, sections through small pieces of the cotyledons of a bean or pea seed : mount on a slide in a drop of iodine solution. Observe that, lying among the starch grains (stained blue by the reagent), there are many much smaller grains which show the yellow-brown colour-reaction characteristic of proteins.   Similarly, cut sections of the dry endosperm of wheat or maize or other grass grains.   Mount the sections in iodine solution, and note that the protein grains or, as they are sometimes called, aleurone grains are confined to the layer just beneath the seed coat. Apply the biuret test to other unstained sections by adding the copper sulphate and potash to the sections placed on a glass slide.   Heat, and, when cool, drain away the fluid, add a drop of water, put on a cover glass and examine under the microscope. Treat other fresh sections with Millon's reagent and examine with a lens : note the red colour of the aleurone grain layer.

It is, however, in oily seeds that the protein grains reach their highest development. In order to prevent the solution of the water-soluble parts of the aleurone grains, we adopt the precaution of moistening with oil the razor used for cutting sections of the endosperm of such seeds, e.g. castor-oil and Brazil nut. Sections obtained from these seeds, mounted in oil (e.g. olive oil) and examined microscopically, show numerous, large, oval, transparent aleurone grains, each with a darker granule (the globoid) at one end. If such sections are compared with others cut in water, it will be seen that, in the latter, a part of the contents of the aleurone grains has dissolved, leaving behind a crystal-like body—the crystalloid—which occupies the larger part of the grain. Apply to sections cut with a dry razor the various tests for proteins. The globoid, to which reference has been made, has been proved to contain various mineral substances, e.g. compounds of calcium, magnesium, and phosphorus, and it has also been shown that these substances disappear from the aleurone grains during germination, passing, as there is reason to believe, to the embryo. Hence to our list of reserve food-materials, carbohydrate, fat, and protein must be added mineral substances; though, in most cases, seeds contain reserves of mineral substances in quantities insufficient for the requirements of the seedling.

CHAPTER IV.

THE changes undergone by the reserve food-materials of the seed during germination: the mode of passage of food-materials from the place of storage (endosperm or cotyledons) to the place of consumption (the growing embryo).

WE learned in Chapter III. that even a very small seed contains a certain amount of those substances, starch or oil, etc., which constitute reserve food-materials, and that the larger seeds possess rich stores both of carbohydrates or fatty and of protein (nitrogenous) food-substances accumulated either in the endosperm or in the fleshy cotyledons. We learned further (Exp. 13) that, in the course of germination, as the embryo develops to the seedling, the food-materials disappear from the endosperm or cotyledons. We inferred, therefore, that they are transported thence to the growing parts of the seedling.

This inference receives support from the experiment, which we now perform, of germinating large and small seeds, *e.g.* broad beans and poppies or "Californian poppies" (Eschscholtzia californica) in darkness.

43. Sow the seeds in small pots containing ordinary soil, and place the pots in a light-proof box (Fig. 12) or cupboard. The pots must be taken out of the dark box from time to time, in order that the seedlings may be observed and measured, but they should be replaced immediately after inspection. As the experiment proceeds, we observe that, though the seeds germinate quite well, the seedlings produced in darkness are curious, pale plants, and that, whereas the bean plants live for three, four, or more weeks, the poppy or Eschscholtzia seedlings live only for a limited number of days. In other words, as we should guess from their appearance, both beans and poppies die ultimately of starvation, the latter very soon, the former only after a fairly long interval of time—not indeed, as we determine by observation, till the reserve food-materials originally contained in them have disappeared from the cotyledons.

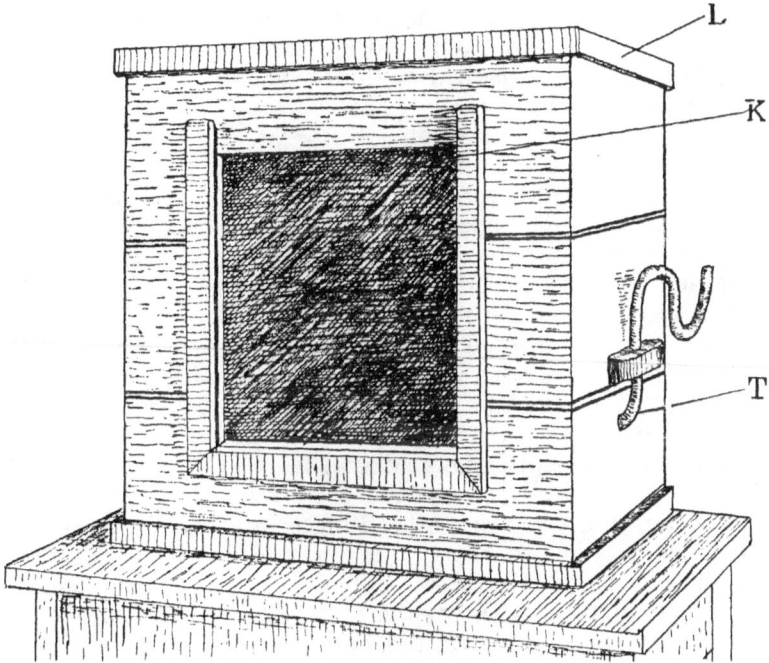

FIG. 12.—DARK BOX. (See Appendix, p. 234.)
L, lid; K, shutter; T, aerating tube.

We cannot fail to attribute the longer life of the dark-grown bean seedlings to the large stores, and the shorter life of the dark-grown poppy or Eschscholtzia seedlings to the slender stores of food-materials which the seeds contain.

But if we recall the conditions in which the reserve food-materials exist in the seed, and the properties of these substances, we find ourselves confronted with a difficulty, namely, that of understanding how these food substances pass from the one place to the other—from the tissues of the endosperm or cotyledons, to the growing root and shoot of the seedling. We know that the starch consists of solid grains which lie in compartments bounded by cellulose walls, we know that starch and oil are insoluble in water, and that much of the protein of the aleurone grains is also insoluble (Exp. 42). Since the reserves are insoluble, it would appear to follow that they are irrevocably imprisoned in the tissues of the endosperm or cotyledons. Yet we have just obtained evidence (Exp. 43) that the reserve-materials are liberated from the tissues and distributed to the embryo. If, however, we state these apparently contradictory facts in a somewhat modified form, we shall see that they are not irreconcilable. The reserve food-materials of the resting seed consist of solid granules (starch and protein) or of droplets of oil: which substances are—except some of the proteins—insoluble in water, and therefore incapable of traversing the cellulose walls of the compartments in which they lie. In the active, germinating seed, food-materials reach the embryo and, at the same time, the reserves in endosperm or cotyledons decrease and tend to disappear. Therefore, the active seed must have a means of so changing the reserve food-materials that they are no longer compelled to remain imprisoned in the tissues in which they were laid down, but are free to travel thence to the embryo. It is evident that, for substances to pass from the tissues of the endosperm to those of the embryo, they must be soluble in water and capable of osmosis. If we accept this reconciliation of our facts, we proceed to ask, into what soluble and diffusible substances may the reserves of the seed be changed? For instance, into what soluble, diffusible substance may starch be transformed? The answer is suggested already by the results of Exps. 18 and 25, which show that starch, boiled with traces of mineral acids, becomes converted into sugar. But if this hydrolysis of starch to sugar actually occurs in the endosperm, it certainly is not due to the agents, heat and traces of mineral acids, employed in our experiment. Granting, therefore, that our reasoning has been sound, we are driven to conclude that there must be present in the germinating seed some special agent which, like the mineral acid, is capable of effecting the hydrolysis of starch to sugar; but which, unlike the trace of mineral acid, is capable of effecting this change with rapidity at ordinary temperatures.   Thus we arrive at a point in our argument at which we formulate a definite hypothesis, and put that hypothesis to the test of experiment.   We state our hypothesis thus : the starch contained in the endosperm of resting seeds is converted during germination, by some unknown agent, into sugar.   To test our hypothesis we ask : (1) does this transformation actually take place? (2) if so, by what agent is it effected?

44.   To obtain a positive answer to the first question we have only to repeat Exp. 24, viz. pound up in a mortar about ¼ of a pound of the grains of sprouting barley or wheat; add a little water, and continue the pounding : filter, and test a portion of the filtrate for sugar (see Exp. 21).

To obtain an answer to the second question, we take the remainder of the filtered extract just obtained, put it in a large beaker or bottle, add gradually, several times its volume of methylated spirit till a heavy precipitate is produced.   Allow the liquid to stand for an hour or more till the precipitate has settled.   Pour off the bulk of the liquid, taking care not to stir the precipitate.   Transfer what remains of the liquid, together with the pre-cipitate, to a filter.   As soon as the liquid has drained away, take the filter paper from the filter funnel, open it out, spread it on a plate, scrape off the precipitate by means of a wooden spatula, and transfer it to a small, wide-mouthed, stoppered bottle.   Add about 20 c.c. of distilled water, shake thoroughly, and filter, if necessary. Put the solution thus obtained into a clean bottle, add a *trace* of an antiseptic, *e.g.* thymol or eucalyptus oil, and label it.   Before investigating the action of this extract on starch, it will be necessary to test a sample of it for sugar. For this purpose take out about 1 c.c., dilute it with 5 c.c. of water, and apply the sugar-test.   If only a slight sugar reaction is obtained, it may be disregarded; but, if the test proves that sugar is present in considerable quantities, the remainder of the extract must be treated again with an excess of alcohol (methylated spirit), the liquid filtered, and the precipitate redissolved in water in the same way as before.

45. Put a very little, pure starch powder in a beaker : add water : boil thoroughly : dilute the boiled liquid until it is almost clear ; take three test tubes labelled A, B, C. To A add some of the starch liquid only, to B starch liquid and about 2 or 3 c.c. of the extract prepared in Exp. 44, to C add water and 2 or 3 c.c. of the same extract. Leave the tubes in a warm place at any temperature between 20°-30° C. At hourly or convenient intervals, test the liquids in A, B, C, for starch, using the method described in Exp. 18. Determine that, whereas the blue starch reaction of the liquid in A remains, the reaction given by B gradually becomes fainter. C, which serves the almost unnecessary purpose of demonstrating that the alcoholic extract itself contains no starch, gives no starch reaction. When B gives only a faint starch reaction, concentrate, by heating in porcelain dishes, each of the liquids in A, B, C, to a small bulk, and test them for sugar. The fact that B gives a well-marked sugar reaction provides us with the proof that our barley or wheat extract contains the substance for which we are seeking, viz. an agent which, at moderate temperatures such as those at which seeds germinate, converts starch into sugar. The experiment proves, moreover, that this agent is extremely potent—for the amount contained in the 2 or 3 c.c. added to B was but little.

46. Whilst the preceding experiment is going on, prepare an extract from *unsprouted* barley or wheat which has been soaked just long enough to become soft. Ascertain, by similar experiments to those made in the previous experiment, that the amount of the starch-hydrolysing agent present in the ungerminated grains is less than in the sprouted grains.

The starch-hydrolysing substance which we have extracted from germinating barley is known as diastase, and may be obtained in commerce. For subsequent experiments, commercial diastase may be used.

It will be interesting to observe diastase at work on the individual starch grains. For this, use commercial diastase (Appendix A) (or our own extract, Exp. 44).

47.* Prepare sections or a fair quantity of scrapings of wheat endosperm : take two watch glasses A and B, with two other watch-glasses to serve as covers. To A, add some of the scrapings or sections and a little diastase solution, to B, add water only. Cover the watch-glasses to check evaporation, and put them in a moderately warm place (temperature 20°-35° C.). After several hours, mount the sections or the isolated grains of the scrapings in

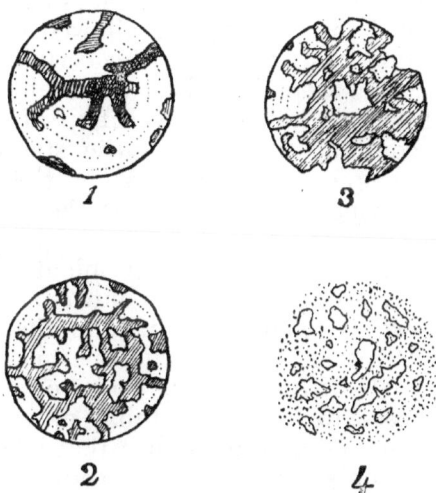

FIG. 13.—STARCH GRAINS OF GERMINATING BARLEY, TO ILLUSTRATE THE ACTION OF DIASTASE.

1, 2, 3, 4. Successive stages of corrosion of the grains. Cf. Fig. 10.
(After Strasburger, *Textbook of Botany*.)

a drop of water, examine them microscopically, and observe that the grains which have been acted on by the diastase begin to show signs of change : they swell, radial fissures appear, each grain comes to have a corroded appearance, and subsequently nothing remains of it but a pale outline or ghost (Fig. 13). Apply the iodine test to the grains in various stages of dissolution. (Avoid in this experiment the use of starch grains, such as those of the bean, which show radial fissures in their natural condition.)

We have thus obtained the experimental proof of the

truth of our hypothesis, and may now, therefore, state it as a fact that the germinating grains of barley or wheat contain an agent which converts starch into sugar. By using other seeds, *e.g.* beans, peas, etc., we prove that diastase, the agent of this change, occurs generally in starchy seeds. A visit to a maltster's works will enable us to see in the malting of barley, as carried on as a preliminary to brewing, a large-scale illustration of the action of diastase. In the preparation of malt, barley is spread out on the floor of a room which is kept warm, the grain is moistened thoroughly, and allowed to germinate. When germination has proceeded to a certain extent, the temperature of the room is raised gradually to a point at which the barley is killed. The action of the diastase is allowed to continue till most of the starch has been converted into sugar, which serves, in the subsequent operation of brewing (see p. 58), for the production of alcohol.

Having proved that, during the germination of starchy seeds, insoluble starch is converted into soluble and diffusible sugar, we conclude that it is in the latter form that carbohydrates migrate from endosperm or cotyledons to the several parts of the growing embryo.

We now proceed to learn something more of the properties of diastase and determine first that it is capable of hydrolysing, not only the starch of seeds, but also that which occurs in mature plants.

48.* This we do by treating with diastase, in the manner described in Exp. 47, sections or scrapings of the flesh of growing potato tubers. After a certain time, the starch grains show corrosion figures similar to those which we have already seen in germinating seeds.

We note, moreover, that diastase has no solvent action on the cellulose of the cell walls which are to be seen in the sections. The starch grains contained in leaves may be shown, by means of sections of leaves treated with diastase, to disappear under the action of this agent. Thus we reach the conclusion that whenever the plant needs to make use of starch, it first hydrolyses it to sugar by means of diastase. We may say that starch is the reserve, or storage form, and sugar the translocation (travelling) form assumed by the carbohydrate material of a great number of plants. Before such a carbohydrate as starch—and the same applies to reserve-cellulose (Exp. 30)—is used as food-material by the plant it is converted into sugar. Hence we recognise that diastase must be as widely distributed in plant-tissues as starch itself.

Now we have used more than once what we know to occur in animals as a starting point for our enquiries into the processes which take place in plants. We will here reverse the procedure and ask: has what we have learned concerning diastase and its action on starch any bearing on animal physiology? We know that starch enters very largely into the composition of animal food, and that the food which we swallow does not pass directly into the body, but into a tube—the alimentary canal—the parts of which are œsophagus, stomach, and intestines. We know further, that the food undergoes digestion, and that not till it has been digested does it pass through the wall of the alimentary canal into the blood. We suspect, therefore, that a knowledge of the action of diastase on starch may help us to an understanding of the significance of the process of digestion which food-substances, such as starch, undergo in the alimentary canal of animals. Just as insoluble starch cannot pass across the walls of the endosperm tissue, so it cannot pass through the complex wall of the stomach or small intestine; and just as soluble, diffusible sugar passes across the cell-walls of the plant, so it may pass across the walls of the absorptive part of the alimentary canal and of the blood vessels, etc., and thus, reaching the blo d stream, be distributed throughout the body of the animal. It will be interesting, therefore, to discover whether diastase, which is present so generally in plants, occurs also in animals.

49. We do this in the following ways. Prepare a solu‧tion of saliva. By chewing a piece of indiarubber or by inhaling a little vapour of glacial acetic acid, a flow of saliva is induced: collect the saliva in a small beaker or porcelain dish: add an equal volume of water: stir thoroughly, and filter. Using the method described in Exp. 45 demonstrate the diastatic action of saliva, i.e. its power to hydrolyse starch to sugar. A sample of commercial, salivary diastase, known as ptyalin (Appendix A), should be obtained and its starch-hydrolysing properties tested.

Food is retained in the mouth for such a short time that much of the starchy material which it contains escapes the action of the diastase (ptyalin) of saliva ; but on reaching the small intestine, the partially digested food is acted on by pancreatic juice, which also contains a diastase. By the action of the pancreatic diastase, the conversion of starch into sugar, which was begun in the mouth, is completed.

The diastase contained in pancreatic juice is present also in the extract of the pancreas, known commercially as liquor pancreatici (Appendix A), and the hydrolysing effect of this liquid on starch may be demonstrated by the method of Exp. 45.

The above experiments throw a clear light on what is meant by digestion. As far as starch is concerned, digestion means the conversion of this substance, by means of the special agent diastase, into soluble, diffusible sugar. They also provide yet another illustration of the fundamental agreement between the nutritive processes of plants and animals. Both use starch as a food-material, and both prepare this substance for distribution through the tissues in precisely the same way. By similar methods of procedure, to some of which reference will be made subsequently, it may be shown that the digestion of other substances, such as fats and proteins, means, in like manner, the conversion of the fat or protein, each, by means of a special agent, into soluble and diffusible substances. Each of these agents is a specialist; thus, diastase acts on starch and starch only. We want a general name to include these specialists in chemical change. They used to be called unorganised ferments, but are now known as *enzymes*. We may form a rough and unfinished picture of the mode of action of the enzyme, diastase, on the starch molecule thus : imagine it holding out one hand to a molecule of water and the other to one pair of the many $C_6H_{10}O_5$ groups which make up the starch molecule (see p. 33). Swinging its two imaginary hands, diastase brings the two $C_6H_{10}O_5$ molecules and the H_2O molecule together, unites them as $C_{12}H_{22}O_{11}$ (malt sugar), and, having done so, is free to lay hold again of a water molecule and another pair of the $C_6H_{10}O_5$ groups, and to treat them in a similar way. Thus we can picture diastase effecting the piecemeal disintegration of the starch molecule by hydrolysing its constituent groups. In this sense we may describe the action of diastase on starch as clastic (breaking down) and hydrolytic (water-adding). It is by similar clastic and hydrolytic processes that other enzymes act.

To return to the study of the properties of enzymes as illustrated by diastase : as we have already remarked, a little of the enzyme goes a long way in producing its specific change. Nor, having regard to our description of the mode of action of diastase, is this altogether incomprehensible : for, according to that description, the enzyme only plays the part of an intermediary, presenting a molecule of water and two $C_6H_{10}O_5$ molecules to one another. The presentation effected, the enzyme is free to bring about another, similar union between water and two more $C_6H_{10}O_5$ molecules. In Exp. 45 we found that even though the diastase was allowed to act on starch for several days, not all the latter was converted into sugar. That this was due not to the exhaustion of the enzyme but to the conditions of the experiment we demonstrate by repeating the experiment thus :

50. Prepare solutions of diastase and of starch (as in Exps. 44 and 45). Add a known amount of the starch liquor to a large test tube or beaker (A), and a corresponding amount to a parchment tube or diffusion shell (B), as described in Exp. 17. Add a measured quantity of the diastase solution to A and an equal quantity to B. Stand A and B in water kept at a moderate and as uniform a temperature as possible. After some hours, and then at convenient intervals, test A and B for starch. At the end of 12-24 hours replace the water in which the parchment tube dips by fresh water. Determine that, whereas A continues to give some colouration with iodine, B ceases, after a time, to give the iodine reaction. This means that, though some starch remains in the test-tube, all traces of it disappear from the parchment tube. Since the only difference between A and B is that, whilst the sugar formed by the action of the diastase on starch accumulates in A, it passes by osmosis from B into the water surrounding the parchment tube, we conclude that the accumulation of the products of enzyme activity tends to bring the action of that enzyme to a standstill. In other words, enzyme action is inhibited by the products of that action.

By a suitable and easily planned series of experiments we may determine that diastase is very susceptible with respect to its starch-hydrolysing powers to external conditions. Thus diastase acts slowly at a low temperature, more rapidly as the temperature increases to about 60° C., and slower again at yet higher temperatures. Heated to about 85° C. it is "killed," *i.e.* its hydrolysing powers are destroyed.

Again, diastase—like other enzymes—may be shown to be very susceptible to other conditions, *e.g.* acidity or alkalinity of the medium in which it acts.

We will now summarise what we have learned concerning the properties and mode of action of the enzyme, diastase. Diastase is a soluble, indiffusible substance (see Exp. 44) occurring in both plants and animals, among which it is widely distributed. Whenever starch is undergoing digestion, diastase is present as the active agent of that change. Starch is a reserve form of carbohydrate food-material. It is insoluble, and hence, though excellent for storage purposes, is immobile and unusable as food. Sugars are plastic forms of carbohydrate. They are soluble and diffusible, and hence are mobile and usable as food. The rôle of diastase is to convert the reserve carbohydrate (starch) into plastic carbohydrates (sugars). Its mode of action is hydrolytic and clastic. A specialist among chemical agents, it acts exclusively on starch. On that substance its action is potent: given suitable conditions of temperature, slight alkalinity of medium, removal of the product of its activity, a very little diastase suffices to effect the conversion of large quantities of starch to sugar. The action of diastase on the starch contained in starchy seeds begins as soon as the latter are placed under conditions which admit of germination. For further information on the subject of diastase, see Bibliography, 10, 11.

Our study of the action of diastase on starch has served to give us a clear and precise idea of what goes on in the course of the digestion of food by an animal. We have seen that both saliva and pancreatic juice contain diastase, and that both, in consequence, convert starch to sugar. If we extend our ideas gained from the mode of action of diastase on starch to digestion in general, we are led to conclude that digestion in animals means—apart from any mechanical action—the production, by the action of special enzymes, of soluble, diffusible, plastic substances, from insoluble, indiffusible materials supplied to the animal in the food. Similarly, applying the same conclusion to plants, we predict that just as a diastatic enzyme converts insoluble starch into soluble, diffusible sugar, so, in oily seeds, a specific enzyme effects the conversion of the oil into soluble plastic substances, and in seeds containing protein (aleurone) grains, other enzymes effect similar changes in the indiffusible and, for the most part, insoluble proteins. Not only shall we expect to find these enzymes in the germinating seeds, but also in any parts of the plants which contain reserves of food-materials liable to be drawn upon for the purpose of nutrition. This hypothesis we proceed to put to the touch of experiment.

51.* Prepare solutions of commercial diastase and of liquor pancreatici (Appendix A). Cut out small cubes of the white of a hard-boiled egg which, as we know, contains insoluble, coagulated proteins. Place a cube in each of two test tubes, and also in a diffusion shell or parchment tube standing in water, add diastase solution to one test tube and diluted liquor pancreatici to the other test tube, and also to the diffusion shell; add a drop of thymol to each of the tubes to prevent the growth of bacteria, and keep them at a temperature of 35°-40° C. We note after some hours that, whereas the white of egg to which diastase has been added undergoes no change, that exposed to the action of liquor pancreatici becomes irregular in outline, smaller and translucent. Noting these changes, we recall the fact already discovered, that protein-like substances (peptones) exist which are both soluble and diffusible (see p. 41, Chapter III.). Now, from our study of diastase, we should not be unprepared to find that if pancreatic juice digests proteins, soluble and diffusible bodies result from this digestion.

52. Therefore, after the pancreatic juice has acted on the white of egg for some time take a sample of the liquid and test it for peptones by the biuret test. The rose colour which results from the application of the test indicates that the pancreatic juice has effected the conversion of insoluble, indiffusible, coagulated albumin into soluble, diffusible peptone. After 12-24 hours, apply the peptone test to a sample of the liquid in the vessel into which the diffusion shell (or parchment tube) dips. If no peptone reaction is produced, concentrate half of this liquid to a small bulk and repeat the test. Note that, though the liquid is shown to contain the protein-like body, peptone, no coagulation occurred during the operation of boiling down the liquid to a small bulk; for peptones, unlike the soluble proteins, are not coagulated by heat. It may be necessary to allow the experiment to continue some days before peptones are recognisable in the water of the outer vessel.

53. Start another digestion experiment with white of egg, or curdled milk, and liquor pancreatici, using, however, a *very small quantity* of the substance to be digested. Before all the solid white of egg or curd of milk has disappeared from the test tube, test for peptone. Then replace the tube and test again at intervals of six or more hours. It will be found that, as the action of the pancreatic juice continues, the peptone reaction gets fainter, and finally fails. Peptones, therefore, are not the final products of the digestion of proteins by pancreatic juice. By appropriate experiments it may be shown that the peptones give place to simpler nitrogen-containing bodies, among which may be mentioned leucin, tyrosin, and asparagin (Bibliography, 11). It would take up too much time to investigate these final products of the pancreatic digestion of proteins; but we may state that they have been shown to belong to certain groups of nitrogen-containing, organic substances known as the amino-acids and their derivatives (Bibliography, 11).

The main facts which we have learned from the study of the pancreatic digestion of proteins are as follows. During digestion, soluble and diffusible substances are produced. The digestion takes place in two stages; in the first stage, peptones are the chief bodies formed; in the second, the peptones are converted into yet simpler bodies. We may term the enzymes which effect these changes *proteases,* though, as a matter of fact, the work is done in stages by distinct enzymes, one of which acts on proteins, and the other on the peptones to which the proteins give rise. By methods similar to those used for the preparation of diastase, proteases may also be extracted from plants, *e.g.* from germinating seeds, and it can be shown that, as the result of their action, the protein reserves contained, for example, in the aleurone grains are converted into soluble, diffusible substances. We will content ourselves with proving that aleurone grains undergo dissolution when acted on by protein-digesting enzymes. For this purpose we use a solution of pepsin (Appendix A) instead of an extract of the seeds containing the grains :

54.* Cut thin sections of the endosperm of a castor-oil seed, using, for the purpose, a dry razor. Mount some in oil, some in water, and place others in watch-glasses containing pepsin. Set the watch-glasses in a warm place. Whereas, after twenty-four hours, the water has dissolved only part of each grain, the pepsin has dissolved the aleurone grains much more completely.

By the use of methods like those employed in the present chapter, it has been proved that the oily and fatty reserves of seeds (and the fats and oils of the food of animals) are hydrolysed by a fat-splitting enzyme (lipase), and yield fatty acids and glycerine.

Hence, we conclude that plants, by means of appropriate enzymes, convert the reserve materials contained in their seeds (and in storage organs generally) into soluble, diffusible substances, and that animals effect, by similar agents, precisely similar changes in the carbohydrate, fatty, and protein materials of their food. In both cases, these soluble, diffusible products of enzyme activity serve the purpose of nutrition. Thus we justify our hypothesis that the food-materials of plants and animals consist of identical substances.

CHAPTER V.

THE meaning of the term Nutrition: the use which the plant makes of food-substances. The germinating seed considered as a machine. The source of the power which drives the machine and the conditions under which it works.

WE concluded from the results of our experiments in Chapter IV. that the reserve materials of the seed are rendered soluble and diffusible by the action of enzymes, and that the plastic food-substances thus produced serve to nourish the seedling. We have now to discover what is meant by the nourishment of the seedling: in other words, we have to find out to what uses the plant puts plastic food-materials.

It would be possible to begin this work by experimenting with seedlings; but our task will be simplified if we choose, not one of the more complex, higher plants, but one of the simpler, lower plants.

Of the lower plants which may be used for a study of nutrition, the most convenient is that used by brewers to set up alcoholic fermentation in sugar. The plant in question is known as yeast (Saccharomyces cerevisiæ), and may be obtained in commerce.

55. Having purchased a small quantity of active yeast, we prepare a nutrient solution, *e.g.* Pasteur's nutrient solution (Appendix A), in which it is known that yeast grows well. Half fill a small flask with the solution; boil it, remove the flask from the bunsen flame, plug the neck with a wad of cotton wool, and set it aside to cool. Pass a needle mounted in a pen-holder once or twice through the flame of a bunsen burner in order to sterilise

its point. Allow it to cool, take up a little of the yeast
on the point of the needle and transfer it to the cooled
liquid in the flask. Plug the neck of the flask again with
fresh cotton wool, label it appropriately, and stand it in
a warm place (25°-30° C.). After some days, note that
the contents of the flask have become turbid. By mounting
a drop of the liquid on a slide and examining it under
the high power of the microscope, it will be seen that the
turbidity is due to the presence of innumerable spherical or
oval bodies, which, if the magnification is sufficiently great,
present the appearance shown in Fig. 14.

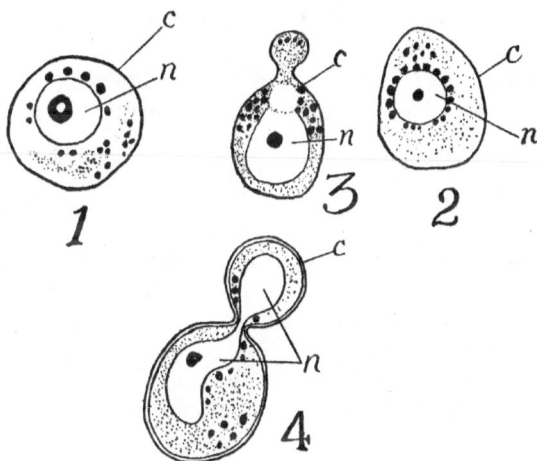

FIG. 14.—YEAST CELLS.

1, 2. Living yeast cells, from a sample of brewers' yeast. Very highly magnified.
3, 4. Living yeast cells, growing actively (budding) in Pasteur's nutrient solution.
c, cell wall; n, nucleus. (After Wager, 1910.)

We note that these oval bodies are of different sizes,
and that some of them are in process of division or of
budding off cells like themselves (Fig. 14). Hence they
are evidently alive. Each oval or spherical particle which
constitutes a complete yeast plant is called a cell or protoplast, and since a single protoplast is capable of nourishing
itself, growing, dividing, and, in fact, leading an independent existence, we say that yeast is a unicellular plant.

56. If a drop of potassium-iodide iodine solution is run in under the cover glass, the granules become of a yellow-brown colour. Similarly, if some of the turbid liquid is filtered through glass wool—or if some of the original yeast is taken and pounded up with clean silver sand in a mortar, it may be shown, *e.g.* by Millon's reagent (Exp. 33), that the pounded contents of the yeast cells contain proteins. We thus confirm the result of the iodine test applied to the individual yeast cells. We conclude that each cell which, since it feeds, grows and divides, is alive, contains substances which, after being killed, give the reactions characteristic of proteins. Since the amount of yeast borne on the needle point has sufficed to produce the relatively enormous quantity now present in the flask, it is clear that there has been a very considerable augmentation of the amount of proteins, etc., present in the original cells placed in the flask. Therefore, the living yeast cell, supplied with appropriate nutrient materials, which include sugar and also traces of nitrogen-containing bodies, builds up proteins and incorporates them into its living substance. In this process of incorporation, both carbohydrate and nitrogen-containing materials are employed. As a result of this process, the amount of living substance is increased. Of the chemical constitution of living substance we know little or nothing. We know that when it is killed it gives invariably protein reactions, and hence infer that proteins are among its essential constituents. We call the living substance protoplasm. We conclude, therefore, that the most important use to which the yeast plant puts the materials which it derives from the nutrient solution is the manufacture of fresh living substance.

Applying these conclusions to the seedlings of flowering plants, we may conclude that one use of the plastic food-substances is to supply the protoplasm of the seedling with materials for the construction of new protoplasm. That this process of manufacture goes on in the tissues of the seedling we already have evidence. We know that during germination the embryo increases enormously in size. We find also, from a microscopic examination of sections through the young root and shoot of seedlings, that their tissues are made up of minute masses of substance, giving, like the yeast cell, the yellow, iodine reaction. The number of these masses is far greater in the germinating seedling than in the embryo contained in the seed. Therefore, we conclude that, just as the yeast plant consists of protoplasts which undergo division, so the tissues of the seedling consist of protoplasts which undergo division. The only essential point of difference lies in this, that, whereas the yeast cells, after division, separate from one another and constitute each a unicellular plant, the protoplasts constituting the tissues of the seedling do not separate, but remain connected with one another. In other words, the seedling is a multicellular organism and not a unicellular organism like yeast.

When we examined the sections of plant-tissues (Exp. 31*), we found that the tissues are divided by walls of cellulose into small compartments, and we recognise now that the protoplasm which gave the yellow-brown colouration with potassium-iodide iodine solution is the living thing—the protoplast, and that the cellulose walls are but a scaffolding or supporting framework to the protoplasts. Since, as the seedling grows, the number of cells increases, there must be, beside the increase of living protoplasm, an increase of the cellulose substances which form the walls around each protoplast. These walls are formed by and from the protoplasm. Part of the latter, undergoing breaking-down processes, produces cellulose, which is deposited either between two cells which have just divided, or on the already formed walls of young cells. Thus, the protoplasm makes from the plastic food-substances, not only more or less permanent additions to its own substance, but also uses up some of its substance in the production of cell-walls. Such manufacture by the protoplasm of the various substances needed by the organism for special purposes is called secretion. We may say that the cell-wall substance is secreted by the protoplasm, that diastase or any other enzyme, such as we know to be produced by plants or animals, is secreted by the protoplasm, and so on. The plastic food-substances supplied by the endosperm or cotyledons to the growing seedling provide materials which serve for the increase of the protoplasm of the seedling.

But not all this increase is permanent, for some of the protoplasm is devoted to the secretion of special substances, such as cellulose, enzymes, etc. Hence some of the constituent molecules of the plastic food-substances migrating from the endosperm become, in the seedling, integral parts of protoplasm; and some, having been for a time constituents of the protoplasm, serve for the formation of the various secretions. Moreover, if the rate of passage of sugar from endosperm to seedling is faster than the rate at which the sugar is being used by the seedling, the surplus may be reconverted into starch and stored temporarily in the seedling in that form (see Exp. 19).

We thus obtain a definite idea of what is meant by nutrition. This process involves the incorporation of certain of the constituents of plastic food-substances with the protoplasm, which becomes thereby increased in quantity; it means also the provision of materials not only for this permanent increase, but also for the manufacture, by the living protoplasm, of the many various substances which play a part in its life.

Now, if this were a complete account of the fate of the plastic food-materials which the seedling derives from its reserve food-materials, it would mean only that the elements contained in the reserve materials pass to the seedling, and, undergoing recombinations, become parts of the protoplasm, of the cell-wall, of enzymes, and of other secretions. If this were the case, there would be, as it were, merely a transmigration of matter from the endosperm or cotyledons to the seedling, and the dry weight of a seedling which has access to the reserve-materials only for its supplies of food, would neither increase, nor decrease, as germination proceeded. Whether this is the case or not we can put to the proof of experiment.

The problem which we have to solve is as follows: If a seed is germinated under such conditions that it cannot obtain supplies of food-materials except from its reserves in cotyledons or endosperm, does the dry weight of the seedling remain constant, or is there an increase or decrease in dry weight?

We know that the germinating seedling takes up water, and, therefore, its undoubted increase in weight might be due merely to the water which it absorbs.   We know also that the cotyledons, *e.g.* of the bean, shrivel as they lose their reserves of starch, etc.   Is the loss of dry weight by the cotyledons exactly balanced by the gain in dry weight by the rest of the seedling?

This is a matter of such importance that though the experiment is tedious, and requires fairly accurate weighing by means of a chemical balance (Appendix B), all students should attempt it.

To carry out the experiment we proceed as follows :

57. Pick out about 60 seeds of kidney bean (Phaseolus vulgaris) or other plant with large, epigean cotyledons. Soak the seeds for twenty-four hours in tepid water.   Select 50 seeds of approximately similar size, divide them into two lots of 20 each, reserving the other 10.   Weigh the two lots of 20.   Plant one lot in pots in moss or coco-fibre, and plant also the 10 reserved seeds in case any of the 20 fail.   Put the pots in a dark room or box, being careful that no light falls upon them.   Give water when necessary.

Squeeze out each seed of the other lot of 20 from its seed-coat, pound the seeds thoroughly in a mortar, and, when they are thoroughly crushed, transfer them to a weighed porcelain dish and dry in a drying oven at 100° C. After a day, take the dish from the oven, place it in a desiccator to cool; weigh, replace in the oven for twenty-four hours, then remove, cool, and weigh as before.   We thus obtain the dry weight of ungerminated seeds (less the weight of the seed-coats, which does not concern us). After a fortnight, when the dark-grown seedlings have grown to a considerable size, turn them out of their pots, remove, by washing, all pieces of moss, etc., from their roots, select 20 seedlings, weigh, chop, and then grind them in a mortar and determine their dry weight.

The difference between the fresh weight and dry weight of the seedlings gives a measure of the amount of water taken up in the course of germination.   A comparison of the dry weight of the 20 germinated seeds with the 20 similar, ungerminated seeds, shows that in the course of germination the dry weight of the seedlings has decreased.

We prove, thus, that during the work of germination

there is an actual loss of material by the plant. Not only is food-material passed from cotyledon to seedling, but also the plant, if prevented from getting food-material from other sources, loses some of the material which it contained before germination began.

In order to form an idea as to the meaning of this loss, let us imagine that we are engaged, not in the study of the physiology of plants, but in that of the physiology of motor-cars; and let us ask what we should require to learn before we could claim to understand the mechanism of cars. We should have to find out what uses the several parts serve, *i.e.* the functions of those parts, and also how they contribute to the work of the car as a whole. But we should require to know more than this; namely, the source of the energy which serves to drive the mechanism. If only from the smell left in their wake, we are familiar already with the fact that most cars consume petrol. We know also that the petrol poured into the tank of the car has to be replenished at intervals, and that, the faster the car is driven, the quicker is the petrol consumed. We conclude that the petrol furnishes the energy which drives the car. We find in the car arrangements for heating the petrol and for mixing it with the oxygen of the air so as to cause it to undergo combustion : that is, to unite chemically with the oxygen. As a result of this combustion, the petrol disappears, being replaced by waste, oxidised substances formed by the union of the constituent elements of petrol with oxygen. It is not, however, to produce these waste substances that the motorist purchases petrol. The waste substances are the unavoidable consequences of the combustion, and have to be got rid of. It is the energy liberated during the combustion that the motorist requires and employs, by means of appropriate mechanical devices, to do the work of propelling his car. If we burn a little petrol in a lamp, we find that energy, in the form of heat and light, is produced, that the petrol disappears, and that gases, carbon dioxide and water vapour, are given off. If we shut off the supply of oxygen completely, the petrol fails to burn; if we reduce the supply of oxygen, it burns with a smoky flame—the combustion is incomplete —and some of the carbon of the petrol, instead of uniting with oxygen to form carbon dioxide, is deposited as soot. The energy liberated during the combustion of the petrol in the lamp may be caused to do mechanical work; to convert water into steam, to make the heavy lid of a kettle containing water heated over the flame of the lamp to bob up and down : so also the energy liberated during the combustion of the petrol in the car is caused to do the work of propulsion.

Now petrol consists of a mixture of substances having the general formula, C_nH_{2n+2}, and when it is burned completely it gives rise to water and carbon dioxide. Moreover, it can be proved experimentally that a definite amount of petrol unites with a definite amount of oxygen, and, as the result of its complete combustion to carbon dioxide and water, liberates a definite amount of energy. Further, whether the oxidation is slow and carried out in several stages, or whether it is rapid and carried out apparently in one stage, the amount of energy liberated is the same. Just as we pour a pint of water from a pot, either in a rush or drop by drop, and in either case get no more and no less than a pint, so the energy liberated by the oxidation of a given amount of a substance is fixed in amount, no matter whether it is liberated in one or more stages.

If we turn our attention again to the seedling, we recognise as we note its root penetrating into the soil, its stem lifting itself into the air, its protoplasts dividing and laying down new partitions of cell-wall substance between the daughter protoplasts, that the plant, like the motor in action, is doing work. But to do work is to expend energy. From what source does the plant derive the energy necessary to carry on the work of living? The loss in dry weight undergone by the seed during germination now becomes significant. The liberation of energy by the combustion of petrol in the motor-car, or in the lamp, involves a consumption of petrol and of oxygen. The activity of the germinating seed involves a loss of substance. May not this loss of substance be associated with the liberation of the energy by means of which the plant does its work? And further, from the analogy with the working of the car, may not this energy be derived from the union of certain substances contained in the seed with oxygen obtained from the air? Thus we arrive at an hypothesis which is capable of verification, and which, moreover, does no violence to our knowledge of seedlings. For we know already that a loss of substance, analogous to the loss of petrol in the running car, occurs in the germinating seed. We know, moreover, that seeds contain substances capable of uniting readily with oxygen : for example, the oil of oily seeds may be used for lighting purposes. A handful of sifted sugar thrown on a smouldering fire burns with a bright flame. It seems reasonable to suppose, therefore, that the reserve food-materials supply the materials for the process, akin to combustion, in the course of which energy is furnished to the plant. But if this, or something like this, is the mode whereby the plant derives energy, we shall expect to find that in doing so it consumes oxygen, just as the petrol-engine consumes oxygen. Thus, we may test our hypothesis by determining the relation of seeds to oxygen. If the seed depends for its supply of energy on the union of certain combustible materials with oxygen, and if the energy so derived gives the seedling its power of doing work, then, just as without oxygen, the petrol will not burn and the car refuses to run, so, without oxygen, seeds should refuse to germinate. To determine whether or no this is the case, we proceed as follows :—

58. Obtain seeds which are capable of germinating under water—rice is good for the purpose, but not always easy to get in a living state : failing the seeds of water- or marsh-plants, onion seed will serve, though it only germinates so far under water as to show the white tip of its radicle. Next, prepare a quantity of boiled water by the method described in the Appendix B. Into each of two similar small flasks or beakers, one filled with unboiled, the other with boiled (and cooled) water, put a dozen grains of rice or other seeds known to germinate under water. Pour carefully a thin layer of oil on the surface of the boiled water in order to check absorption of oxygen from the air. Observe the seeds at intervals, and note that germination does not occur in the absence of oxygen.

The following more striking method of illustrating this dependence of germination on oxygen-supply is due to Dr. F. F. Blackman :—

59. Two large beakers are placed in a sink; soaked barley grains (about a dozen) are put in each, and a wide-mouthed funnel is inverted over the grains in each beaker. The beakers are filled with water so that the grains are submerged. One beaker is put under an open tap so that a constant stream of water passes down the neck of the funnel and overflows from the top of the beaker.

To the other beaker, no fresh water is supplied. Hence the grains in the first beaker receive a constant supply of oxygen dissolved in the water of the stream and those in the other receive only such supplies of oxygen as are contained in the water in the beaker, or are absorbed by the water from the air. In the course of a week or so, the barley in the continuous water-stream, although submerged, germinates and puts forth green leaves which grow to a considerable size; that in the standing water, though germination may begin, is unable, in the absence of sufficient oxygen, to produce any but the most diminutive of seedlings.

In addition to these methods, the following experiments may be employed to demonstrate that oxygen is necessary for germination :—

60. Fill two gas jars with water and invert them so that their open ends are downward and beneath the surface of the water in a pneumatic trough or large dish. Displace the water by carbon dioxide (Appendix B). Impale two soaked bean seeds on a long blanket- or hat-pin. Fix the pin to the inner surface of a cork which fits the mouth of the jar, and then push the cork with pin and seeds into the open end of one of the gas jars, keeping the lower end of the jar under water during the operation. Proceed in a similar manner with the second jar, but before pushing its cork home, introduce, by means of a pipette, a little air into the jar.

Leave the jars standing in the water in the dish or pneumatic trough to prevent any air from entering; and proceed in a similar manner to prepare two jars containing each, two soaked, impaled bean seeds. Replace the water in each jar by hydrogen (Appendix B), and, having done so, allow a little air to enter one of the jars; since, however, hydrogen and oxygen form, in certain proportions, an explosive mixture, the hydrogen experiment should be omitted by those unaccustomed to chemical manipulation. Instead of hydrogen, nitrogen gas (Appendix B) may be employed.

As a contrast, set up a similar gas jar containing air and pass soaked beans, transfixed by a blanket-pin, into the jar. Record from day to day the germination of the seeds in the jars. From the fact that germination proceeds fairly uniformly in all the jars containing air, we infer that the various gases (CO_2, H_2, etc.) do not of themselves affect germination adversely; and since it proceeds not at all, or at most very slightly, in the jars from which oxygen is absent, we conclude that, as our hypothesis led us to expect, oxygen is necessary for germination.

Another method which, though less direct, is interesting because of its bearing on garden-practice, may be tried.

61. Prepare enough puddled clay, by kneading stiff clay with water, to fill two fair-sized marmalade pots. Put a layer of the wet clay at the bottom of one pot, plant a dozen barley grains, and press wet clay closely over them till the pot is full. Cover the pot with a sheet of glass. Fill the second pot with puddled clay, and plant barley grains just below the surface. Cover each pot with a glass plate. After a week or more, investigate the germination of the barley in the two pots. (The puddled clay must be prevented from drying during the course of the experiment, otherwise it will crack, and the experiment will fail. The drying may be prevented by covering the two pots with a bell-jar, under which a saucer of water is placed.)

The results of the preceding experiments show that there can be no doubt that oxygen is necessary for germination. If we succeed in establishing our hypothesis, we shall have the right to conclude that the failure of seeds to germinate in the absence of oxygen is due to the fact that, no combination between oxygen and combustible substances being possible in these circumstances, no energy derived from this source is available, and no work can be carried on. But if oxygen is used for this purpose, the air in which seedlings are growing must in consequence come to contain less oxygen. By taking advantage of the well-known fact that a taper burns in ordinary air, but not in air free from oxygen, we can determine whether germinating seeds extract oxygen from the air.

62. Half fill each of two wide-mouthed, well-stoppered bottles with soaked peas or barley. Replace the stoppers and put the bottles in the dark in a warm place. After twenty-four or forty-eight hours, bring *one* bottle into the room. Having lit a taper, remove the stopper, and plunge the lighted taper into the air of the bottle. The fact that the flame is extinguished gives us proof that oxygen has disappeared. Since oxygen has disappeared from the air in which the seeds have germinated, it is in a high degree probable that it has entered into combination with some oxidisable substance contained in the seeds. We will assume that this has been the case and also that the oxidisable substance is sugar. Now a sugar, when burned, gives rise to carbon dioxide and water according to the equation :—

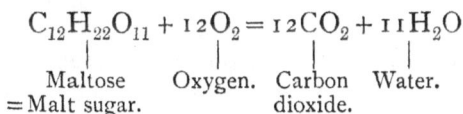

$$C_{12}H_{22}O_{11} + 12O_2 = 12CO_2 + 11H_2O$$

| Maltose | Oxygen. | Carbon | Water. |
| = Malt sugar. | | dioxide. | |

Therefore, by ascertaining whether carbon dioxide is produced by germinating seeds, we have a method of testing the truth of the assumption which we have just made. For the purpose of this test, we prepare a solution of lime-water or of baryta-water (see Appendix A), pass into a sample of either of these substances a stream of carbon dioxide from a CO_2-generating apparatus (Appendix B), and note that the lime-water (or baryta-water) becomes cloudy. The cloudiness is due to the production of calcium (or barium) carbonate :—

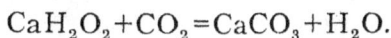

$$CaH_2O_2 + CO_2 = CaCO_3 + H_2O.$$

By blowing through a glass tube, dipping into another sample of lime- or baryta-water in a test tube demonstrate that the air expired from the lungs is rich in carbon dioxide. We now apply the test for CO_2 to the air in the second bottle of seeds which has remained in darkness : remove the stopper, pour in a quantity of lime-water, replace the stopper, and shake the bottle. The marked turbidity of the liquid indicates that carbon dioxide (CO_2) was present in the air of the bottle in considerable quantity. Since, however, ordinary air contains CO_2, it will be well to test a sample of air thus: empty out the bottle, wash it, and then pour in lime- or baryta-water, replace the stopper, shake, and note that the amount of CO_2 enclosed in the bottle is sufficient to produce, not a heavy precipitate like that obtained before, but merely a faint cloudiness.

The consumption of oxygen and the production of carbon dioxide by germinating seeds may be demonstrated in various other ways, e.g. :—

63. (1) Prepare an apparatus (see Fig. 15) consisting of a glass flask (A) fitted with a rubber cork. By means of stout rubber tubing connect the side-tube of the flask with a glass tube of small bore (B), bent at right angles. Introduce into the flask a piece of moist blotting-paper and also one or two dozen barley grains which are beginning to germinate. Stand in the flask a tube containing a strong solution of potash (C), to absorb the carbon dioxide. Replace the cork, warm the flask by dipping for a minute the lower part in water heated to about 40° C. : and place it so that the open end of the bent tube dips below the surface of mercury contained in a small vessel (D). As the air in the flask cools, it contracts, and the mercury rises up the tube to a certain height. Mark the height to which it rises, either by making a line with india-ink on the tube or by reading off the level on a millimetre scale fixed behind the tube. Cover the whole apparatus with a box or black cloth to exclude the light, and, at intervals, record the rise of the mercury in the tube. The rise indicates that the total volume of gas in the flask and tube decreases. We know from the preceding experiments that oxygen is used, and that carbon dioxide is produced by germinating seeds : since the carbon dioxide is absorbed by the potash, the extent of the rise of the mercury in the tube gives an indication of the amount of oxygen consumed by the germinating seeds.

Let us now review the situation to which this series of

experiments has led us. We know that a germinating
seed is doing work, and therefore expending energy. We
know that, in machines such as motor-cars, the driving
energy is derived from the oxidation of petrol (or coal,
or similar bodies). We know that, unless oxygen is supplied, the petrol cannot develop energy, nor can the seed
germinate : when oxygen is supplied, the petrol is consumed and the seed loses in dry weight : when the petrol is

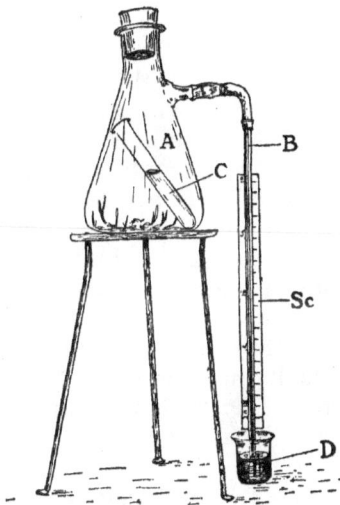

FIG. 15.—APPARATUS TO DEMONSTRATE EVOLUTION OF CARBON
DIOXIDE BY GERMINATING SEEDS.

A, flask, with germinating seeds ; B, glass tube ; C, test-tube containing potash ;
D, mercury ; Sc, scale.

oxidised, waste products, such as carbon dioxide, are liberated. We cannot but conclude that the mode by which
energy is obtained is similar in the motor-car and in the
germinating seed, namely, by the union of oxygen with
combustible materials and the consequent breaking-down
of these relatively complex materials to simpler substances,
such as carbon dioxide and water. Nor is it difficult
to obtain evidence to show that the materials oxidised in
the germinating seed are, as we have assumed, the plastic
food-materials. If we look at the equation on p. 69, show-

ing the mode of oxidation of sugar, we note that the number of molecules of oxygen consumed in oxidising a molecule of malt sugar is equal to the number of molecules of carbon dioxide produced in the process; and, since all molecules occupy the same space, it follows that, in the oxidation of sugar, the volume of oxygen which disappears is equal to the volume of carbon dioxide produced. Thus, we have a means of testing whether, in a starchy seed, a carbohydrate is the material, the oxidation of which liberates the driving energy whereby the plant-machine is kept at work.

We use for the purpose the apparatus of Exp. 63, making, however, the following modifications :—

64. A second flask of similar size and shape is fitted up and placed beside that containing germinating seeds. If no potash is added to either flask, observation shows that the mercury column, though its height may fluctuate owing to temperature-changes in the air, does not rise in the same manner as in the previous experiment. That such fluctuations as occur are to be attributed to changes of temperature causing expansion or contraction of the air in the flasks may be seen by observing that they occur alike in the empty flask and in that containing germinating seeds. Inasmuch, therefore, as the volume of air contained in the flask in which the seeds are germinating undergoes little or no change, we may infer that the amount of oxygen consumed is about equal to the amount of carbon dioxide liberated by the seedlings.

We conclude from the foregoing experiments that, in starchy seeds, the carbohydrates supply the material which, when oxidised, yields the energy whereby the plant does its work. Thus we have satisfied ourselves that the plastic food-substances which reach the embryo from the endosperm or cotyledons not only supply the protoplasm of the embryo with materials out of which is formed more protoplasm, but also serve as "combustible" materials by the oxidation of which the energy necessary for carrying on the life-work of the seedling is liberated. We may speak of the processes whereby the plastic food-materials are produced from the reserve-materials, and whereby they are incorporated with the protoplasm as the nutritive processes;

or we may regard the former, digestive processes, as a preliminary to nutrition, and confine the term nutrition to the latter process only. If we adopt the second course, we require a term to include not only the digestive processes, but also other chemical changes which occur regularly in the living organism. We may use the term *metabolism* in this wide sense. To the process of oxidative decomposition of plastic food-substances—that is, the energy-producing process—the term *respiration* is applied. Since the whole significance of respiration lies in the liberation of energy, it is as absurd to say that respiration consists in the taking in of oxygen and the liberation of carbon dioxide as it is to say that a motor-car is driven by letting oxygen into the engine of the car and letting out carbon dioxide.

That the temperature at which respiration goes on in the seedling is far lower than that at which the petrol is burned in the car need not greatly trouble us. For, in the first place, chemists have shown that chemical reactions generally which go on at a high temperature, may go on, also, though at a slower rate, at a low temperature; and, in the second place, just as we found the enzyme, diastase doing at a low temperature what a trace of mineral acid could do with equal rapidity only at a high temperature, so it may be that enzyme-like bodies are excreted by the protoplasm for the special purpose of speeding up respiration.

There is still one aspect of the production of energy by the motor-car and by the seedling which requires attention. It is well known that the motor-car owes its existence to the improvements which have been made of recent years in the petrol-engine. The petrol-engine has been rendered more efficient by these improvements, that is to say, an increased percentage of the total energy developed by the combustion of the petrol is used for doing the work of propulsion. But, in spite of all the ingenuity expended on the construction of petrol, steam, and other engines, a very large proportion of the total energy produced is inevitably wasted—that is, is not used for the purpose which the engine is designed to serve. This wasted energy takes the form of heat, and the fact that it is wasted is due, of course, to no lack of skill on the part of engineers, but to the properties of matter, *e.g.* to friction and to the power metals have of absorbing, conducting, and radiating heat, and so on. If our comparison of the plant with a machine is just, we shall expect to find that, of the total energy produced in respiration, a large proportion is not employed in driving the plant's machinery, but appears in the form of heat. That this is the case may be shown in a striking manner by the use of the familiar Dewar flasks (Thermo flasks) (Appendix B). As is well known, a hot substance put into a Dewar flask remains hot for a long time, and a cold substance remains cold.

65. Soak in tepid water for twenty-four hours enough peas to nearly fill two Dewar flasks. Wash the seeds in running water, and reject any which seem bad. Remove superfluous water by means of a clean cloth or blotting-paper. Nearly fill one flask with the seeds: whilst filling, insert a small thermometer in the flask so that the part of the stem which records temperatures above 30° C. only remains visible. Push a large wad of dry cotton wool into the neck of the flask to close the neck and hold the thermometer in place. Wrap the whole apparatus in a large, loosely-fitting jacket of cotton wool. Boil the remaining peas, and, adopting similar precautions to those used with the unboiled peas, put half their number in the second Dewar flask with a thermometer and wad and covering of cotton wool. Hang up a third thermometer in an ordinary flask to give the air-temperature during the course of the experiment. During several days, and at as regular intervals as possible, record the temperatures registered by the three thermometers. Observe that the germinating peas produce sufficient heat to raise the contents of the flask to a temperature as much as 10° C., or more, above that of the air. Note that the temperature of the flask containing the boiled peas is for some time little above that of the air; but that, after a day or two, it may also show a rise of some degrees. If it does, turn out the contents of this flask and observe that the peas have become mouldy. Now take the remaining soaked peas and warm them for a few minutes in a beaker of water to which a trace of corrosive sublimate has been added. It should be borne in mind that corrosive sublimate is in the highest degree poisonous, that only a mere trace need be added to the water in the beaker, and that the vessels which have contained it must be thoroughly washed immediately after use. If there is any objection to the use of corrosive sublimate, the peas may be boiled in a solution of permanganate of potash instead. Rinse out the Dewar flask with a solution of corrosive sublimate or permanganate of potash. Transfer the poisoned peas, after wiping them between sheets of blotting-paper, to the clean flask. Arrange the thermometer and cotton wool as before, and determine that no mould being able to grow, no rise of temperature occurs. We learn from these experiments that germinating peas liberate a considerable amount of heat, and conclude that, just as in machines not all the energy is utilised for doing mechanical work, so, in germinating seeds, some of the energy developed in respiration appears in the form of heat, and is therefore not used directly in doing vital work. We learn, incidentally, that low forms of plant-life, such as moulds, also produce heat, and we suspect that they too respire, and that they, like the seedling and the petrol- or steam-engine, are of only limited efficiency. Under more natural conditions than those which obtain in a Dewar flask, the temperature of germinating seeds would, of course, not rise so high, for heat would be lost by conduction to surrounding objects. Although we may regard the heat produced during germination as energy lost, in the sense that it is not applied to the performance of the work of the seedling, yet the heat may be serviceable in other ways, namely, in providing a temperature at which the organism can work at its best. If the petrol in a car is frozen, the car cannot be started; if the temperature of the car rises above a certain point, damage ensues. In other words, the mechanism of the car works only within a certain range of temperature. That this is also the case with the seedling, we have already determined in part : for we know that seeds heated to the boiling point of water are killed.

66. By exposing seed in germinators to different temperatures, e.g. freezing point, room temperature, 25°-30° C., and so on, we may determine that germination proceeds best at a certain (moderately high) temperature. Thus, like machines in general, the plant-mechanism works only under certain definite temperature-conditions.

The facts we have learned respecting the nutrition and respiration of seeds are fundamental facts. They teach us that the living substance (protoplasm) owes its increase to the plastic food-materials, and they explain the source and mode of liberation of the energy whereby the living mechanism works. Therefore, if we hold fast to our general hypothesis which has helped us so much already, and according to which the vital processes of plants and animals are fundamentally similar, we shall expect to find that the modes of nutrition and of respiration of the mature plant and of the animal are those of the germinating seed. The proof that this is the case with respect to nutrition we will defer to Chapter VI., and will deal at once with respiration. With respect to animals, the necessary proofs lie close at hand. We have already traced the plastic food-materials to the blood-stream, and know that the multitudinous protoplasts which make up the body are bathed with fluid derived from that stream. We are aware that the body of the animal is constantly doing work, and therefore expending energy. We know, moreover, that, in the warm-blooded animals, heat is produced in sufficient quantity to maintain the temperature of the body at a remarkably constant level and considerably above that of the surrounding air. We breathe on a frosty day on a piece of glass and note the water which condenses from our breath; we blow through a tube into lime-water and demonstrate, by the precipitate of calcium carbonate which forms, that the air we expire contains carbon dioxide. Putting these facts together, we recognise that the respiration of animals is similar to that of plants. The only apparent difference consists in this, that the waste products of animal metabolism include not only water and carbon dioxide, but also nitrogenous substances—urea, uric acid, etc. Since these substances are evidently derived from proteins, we are bound to infer that, in addition to carbohydrates and fats, proteins also contribute, by their decomposition, to the energy required by the organism for the purposes of its life-work. In point of fact, protein-

decompositions similar to those which take place in animals
are known to occur in plants. We may perhaps express
the facts thus : the bulk of the energy which is used
by plant and animal is derived from the oxidative
decomposition of carbohydrates or fats. In the course
of its activity, the protoplasm itself wears out, and
so produces waste nitrogenous substances. If plentiful
supplies of proteins are at the disposal of the organism,
as is the case with many animals and some plants, the
proteins may be respired, *i.e.* they may be split up, certain
of their products oxidised, and thus release energy. The
only difference between the animal and plant with respect
to these energy-releasing processes appears to be that,
whereas many animals are protein-spendthrifts, most plants
are protein-misers. This is a matter which, though interesting, is not essential, and need not therefore detain us.
With respect to the respiration of the adult plant, we
convince ourselves by the following experiments that what
we have learned from the seedling applies word for word
to the adult.

67. Grow two bean plants, each with its roots in tap-
water contained in a wide-mouthed jar. When they have
passed the seedling stage compare the plants as to root
and shoot development, and then remove one of them from
the jar and place it with its roots in another jar containing
boiled water (Appendix B). A layer of oil is then placed
on the surface of the water in this jar. The growth of
the " transplanted " plant at once falls behind that of the
untouched plant. In the absence of oxygen from the
water the energy necessary for the growth of the root is
not developed. By repeating Exp. 62, using the opening
buds of horse-chestnut or dandelion, etc., we prove that,
like the seedlings, these parts of the mature plant absorb
oxygen and evolve carbon dioxide. By the method of
Exp. 65, using buds or young leaves, we demonstrate that
their respiration results in a production of heat. Thus we
conclude that the energy-liberating processes, which we
include under the term respiration, run the same course in
both young and adult plants, and in animals. We conclude, in fact, that this process is general in all organisms,
and that if there are any organisms in which respiration

does not take place in this way, they must be regarded as constituting special cases, and must receive special consideration.

CHAPTER VI.

THE seedling as an independent plant: the lowest forms of plants and
animals and the lines followed in the evolution of the higher plants
and animals. The distinguishing characters of root- and shoot-systems.
The mode of growth of the root: the functions of its parts: the root-
hairs, the absorbent organs of the root.

THERE comes a time when the seedling, reared hitherto at
the expense of the reserve-materials of the seed, cuts itself
adrift from the latter and sets out on its career as an
independent plant. That career we will now follow. The
independent seedling grows, forms new members, performs
remarkable movements, and becoming a fully grown-up
plant, bears flowers and sets seed. For these operations,
it must be able to obtain large supplies of material, and,
for them, no small amount of energy is required. We have
learned by experiment what substances the young plant and
the animal use in constructing their tissues and how they
obtain their supplies of energy, so, till we find that the
hypothesis is wrong, we will maintain that the mature plant
uses similar food-materials for its nutrition. Since, however, the substances, such as sugar, proteins, etc.,
which, on this hypothesis, constitute the true food-materials
of organisms, are not present in the soil or in the air, it
follows that either our hypothesis is wrong, or that the
mature plant manufactures these plastic food-materials
from other substances which it obtains from the soil or air,
or from both these sources.

68. The following experiment affords evidence in support of our hypothesis :—

Germinate two lots of radish seeds on damp blotting-
paper. When the seedlings have emerged, grow one lot
with the roots in clean tap water, the other lot in tap

water to which has been added a *trace* of potassium nitrate. It will be found that the seedlings with their roots in the water to which potassium nitrate has been added grow far more vigorously than those whose roots are in ordinary water (Fig. 16).

What relation, if any, exists between the absorption of such a mineral salt as potassium nitrate (KNO_3) and the formation of plastic food-materials need not concern us now. We are content to note the evidence supporting the view that the root absorbs substances from the soil, and that these substances contribute in some way to the formation of plastic food-materials on which the growth and activity of the plant depend. Recognising, as we must, the vital importance of plastic food-substances, it will be evident that the first and constant care of the plant must be to obtain adequate supplies of the raw materials which it needs for their construction; and, if we bear in mind what we have learned already with respect to the phenomenon of adaptation—the automatic adjustment of the members of an organism to their special work—(p. 22) we shall expect to find that the plant-members concerned in the absorption of the raw materials of the food show adaptations tending to fit them for this work. That is, these members, or parts of them, present appearances which we may recognise as fitting them to serve as *absorptive organs*. Hence, by a careful and intelligent inspection of the members of a plant (roots, leaves, etc.), we shall obtain broad hints of the special functions which these members or their parts perform.

Our task, therefore, in this and in succeeding chapters is to discover what are the raw materials of the food, how they are absorbed, and how, from the raw materials, the plant manufactures the finished articles, the plastic food-substances such as sugar and proteins, on which the protoplasm subsists and on which, therefore, all growth and activity depend. In studying these problems, we shall learn how profoundly the form characteristic of plants is determined by the necessity under which they labour of producing organs suited for the absorption of the raw materials of the food. We shall even discover that in this necessity lies the origin of the remarkable

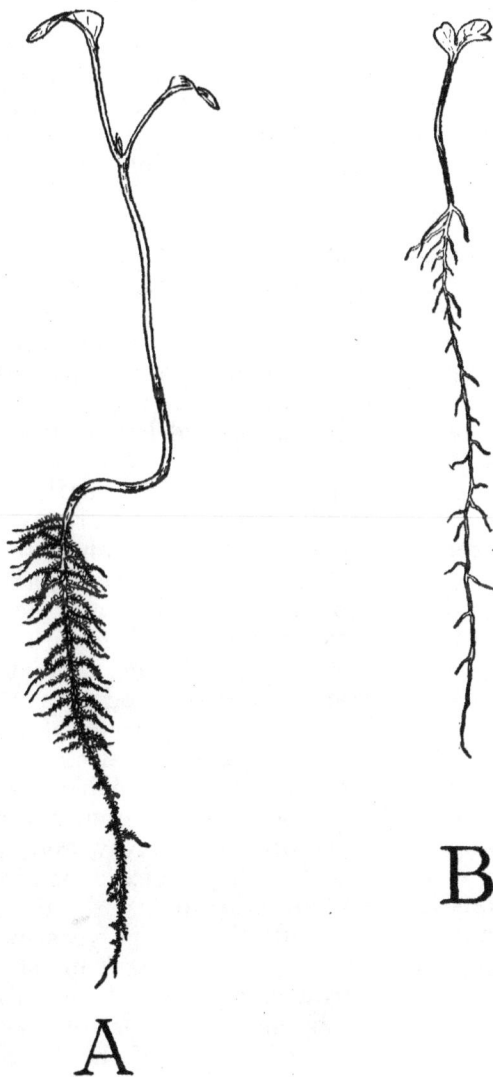

FIG. 16.

A, Radish seedling grown in tap water, to which has been added a trace of potassium nitrate ; B, radish seedling, of the same age as A, grown in tap water only.

differences in form which distinguish the higher plants from the higher animals. Between animals and plants there are, as the evidence of former and later chapters proves, no fundamental differences except those arising from the obligation laid upon the plant to make its own plastic food-materials. The modes of nutrition of the protoplasm are identical in plant and animal : the respiratory processes are similar : the movements, though more conspicuous in the animal, are alike, and the modes of behaviour to external influences are the same. All the wonderful differences in form which stand out when we compare the animal and vegetable kingdoms are the outcome of this great fact that plants make their plastic food-materials, animals appropriate them ready-made. It is a fact of the profoundest physiological importance that "all flesh is grass."

Let us, in order to illustrate these remarks, compare one of the lowest plants with a very simple organism which, though undoubtedly an animal, stands very near the parting of the ways which lead respectively towards the animal and the vegetable kingdoms. The plant may be obtained often in great numbers in ponds, water-butts or rain puddles : the animal is found also in these situations. If water from one of these sources is stood in a glass dish near a window the presence of the plant and of the animal may be revealed by a green scum appearing on the side of the dish near the window. On microscopic examination of a drop of the liquid, minute green organisms in active, dashing movement may be observed. Now and again one of them rests awhile and then may be examined in greater detail. It consists of a single protoplast the greater part of which is not colourless like that of the yeast plant, but green ; the green part having a somewhat cup-shaped form (Fig. 17 A). In the narrow cavity of the "cup," the protoplasm is colourless, and, by suitable modes of preparation, a denser, more or less spherical part of the protoplasm—the nucleus—may be seen lying toward the lower end of this neck of protoplasm. At the free end of the neck is a slight projection or knob, from which extend two protoplasmic threads (flagella), each as long, or longer, than the body. It is by means of the flagella

that the organism rows itself along. On one side of the
body, near the anterior end, a red speck—the eye-spot—
may be seen; and suitably stained microscope-preparations show that the protoplast is enclosed by a delicate wall
of cellulose, through which only the flagella project.   By
hunting through drops of water obtained from puddles of
rain water—and if the student has a microscope at his
disposal the hunt is well worth making—examples of the
second organism may be obtained.   It consists of a single
green protoplast, which has the form of an elongated cup
or flask (Fig. 17 B); but instead of the narrow neck being
occupied by colourless protoplasm, it is hollow, and thus
forms a gullet.   Flagella, attached near the bottom of the
gullet, project beyond the body of the animal.   As in
Chlamydomonas—the plant previously examined—so in
this animal, Euglena viridis, a nucleus and an eye-spot
are present; but, unlike that of the plant, the enclosing
wall of the body of Euglena is not of cellulose.   Upon
occasion, both Chlamydomonas and Euglena may lose their
green colour, become colourless and yet increase in size
and even multiply by division.   So alike are the two
organisms, that it might seem mere hair-splitting to
describe the one as a plant and the other as an animal.
Nevertheless, it is proper and useful to distinguish
them in this manner; for, it is possible to imagine
that, by a long series of changes, all of the nature of so
many adjustments to the more adequate winning of raw
material from water or from land, or to the more efficient
elaboration of plastic food-substances from this raw
material, the higher plants, even the highest flowering
plants, have been derived from a Chlamydomonas-like
ancestor.   So it is conceivable, also, that all animals have
been derived from some lowly form not far removed from
Euglena viridis.   What then is the essentially plant-like
character of Chlamydomonas, and what the essentially
animal-like character of Euglena?   Chlamydomonas encloses its protoplast within a definite wall of cellulose.
Euglena has a sac or gullet open to the outside.   Hence
Euglena can engulf solid particles, digest them, and so
obtain plastic food-substances. Chlamydomonas debars
itself from all solid food supplies; for only substances in

solution can pass its wall. Euglena can, upon occasion,
prey on other organisms, and pick up nutritious morsels.
Chlamydomonas cannot. Chlamydomonas, by surrounding
itself completely by a wall, condemns itself to the labour of
constructing its plastic food-substances from such soluble,
raw materials as its environment provides; Euglena, though

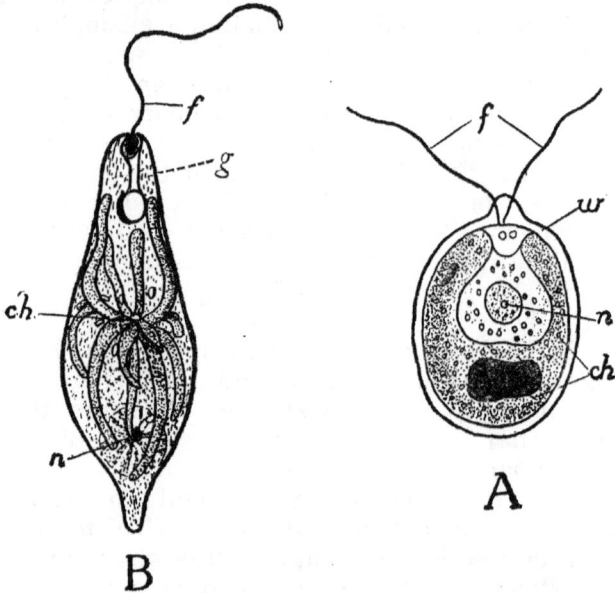

FIG. 17.
A. Chlamydomonas sp. B. Euglena viridis. (Highly magnified.)
w, wall; ch, chloroplast; n, nucleus; f, flagella; g, gullet.

it has not lost this power altogether (see Chapter XI.), can
swallow any likely particle, and thus get its food ready-
made. These are the salient, differentiating characters not
only of Chlamydomonas and of Euglena, but also of plants
and animals in general. It is the investing wall which
determines the mode of life of Chlamydomonas: it is the
gullet which makes Euglena free of all solid food-supplies
which it can engulf and digest. It is the fact that
Chlamydomonas has a continuous solid cell-wall, which is
the essential fact, not that the wall is of cellulose;

though, since it is of cellulose and since animal protoplasts have no cellulose walls, it is sometimes said that the cells of plants are characterised by cellulose walls, and those of animals by the absence thereof. Evidently this is true; but evidently also the wall is the thing, the chemical nature of the wall is a matter of less importance.

We may summarise the whole difference between Chlamydomonas and Euglena thus : Euglena would, if it could swallow it, feed on Chlamydomonas; Chlamydomonas cannot swallow, and, therefore, cannot feed upon Euglena, nor on any other undissolved substance. For " Euglena " write " animals," and for " Chlamydomonas " write " plants," and the statement holds good. Animals feed on plants, plants cannot feed on animals unless, as happens in the case of certain parasites, a plant gets access to the soluble substances contained in the body of an animal.

Our excursion into the world of microscopic plants and animals has been of use, not only in presenting our immediate problems to us more definitely, but also in enabling us to take a more comprehensive view of the plant and animal kingdoms.

We return now to the work of flowering plants, and proceed to enquire into the nature and mode of absorption of the raw materials of their food.

It is a significant fact that plants in general live, as it were, in two worlds. One-half of the plant leads a subterranean life, the other half an aerial life. It is noteworthy, also, that the underground root-system differs markedly in form and aspect from the above-ground shoot-system.

69. A careful examination of a young plant, *e.g.* sunflower or bean, etc., dug up from the soil or turned out from its pot, **or,** better, raised for the purpose in a box of coco-fibre or sawdust (Appendix B), shows us that, whereas the green shoot-system consists of a main axis bearing two kinds of lateral members, namely, lateral axes (branches) and leaves, the colourless, or non-green, root-system consists of a main axis (tap-root) bearing lateral members (lateral roots) of one kind only. Moreover, an examination of the lateral members of the root and shoot shows that they differ from one another with respect to origin, structure, and behaviour. Note, for example, that each lateral shoot begins as a bud standing in the axil of a leaf (see p. 18); that these axillary buds arise as some-what superficial outgrowths from the main axis; that each main and lateral shoot-axis terminates in a bud; and that such a bud may develope into either leaf-bearing or flower-bearing branches; now observe that the lateral roots do not arise from buds, but push themselves out from the deep-lying tissues of the main root, and that, instead of terminating in a bud, each lateral root, like the main-root, has over its tip a thin, brownish, thimble-like covering called a root-cap; nor are flowers borne on the root.

70. Beside observing these morphological differences between the members of the root- and shoot-systems, we prove by the simple experiment of turning a pot-plant on its side and leaving it in a horizontal position for several days that remarkable physiological differences also exist between them. Whereas, in consequence of this change of position, the main stem or ascending axis curves till its young tip points once again vertically upward, the main root (the descending axis) curves in precisely the opposite manner till its tip points vertically downwards.

Whilst drawing the specimen, we note, also, that the lateral members of root- and shoot-systems exhibit char-acteristically different curvatures as the result of this change of position.

When we study the curvatures of root and shoot, we recognise that they serve to bring the root into the soil and the shoot into the air. If we observe an animal making similarly purposeful movements, for in-stance, a worm thrusting its head out of its burrow, we say that the movements are instinctive. Now, in point of fact, certain fixed animals perform movements, in response to change of position, quite like those of the root and shoot of our plant. Hence there is every reason for us to regard the upward and downward bending of shoot and root respectively as examples of nervous activity of the same order as those which in animals are spoken of as instinctive. Like instinctive actions in general, these root- and shoot-curvatures serve evidently to secure the well-being of the organism. The well-being of the plant, then, is secured by the penetration of the root into the soil and by the exposure of the shoot to the air. In other words, the shoot has work to do which can best be done in the air; the root has work to do which it can perform best in the soil.

Thus we come to ask, what is the nature of the work performed by the root-system? The most commonplace observations suffice to convince us that the root-system of a plant serves to fix it firmly in the soil. The tree resists all but the most violent hurricanes; the place on the wall where last year's ivy grew is marked by the remains of the roots, which broke rather than relax their hold when the ivy was torn away; the delicate root of the tender seedling is held so firmly in the soil that the pull of the shoot extricating itself from the seed does not disturb it. We must discover by what peculiar mode of growth the root, though elongating all the while, maintains its hold upon the earth, and so serves as a support to the plant.

71. To this end we choose several germinating bean seeds, the roots of which are about 1-1½ inches long. The seeds for the experiment may be germinated either in moist air, or, since roots often fail to grow straight in air, in boxes of moistened coco-fibre or of sawdust. Dry with blotting-paper the surface of a straight root about 1-1½ inches long, and, by means of the method given in Appendix B, make a series of 10-12 india-ink marks at distances of about 1 mm. from one another, starting from the tip of the root. Lay a millimetre scale beside the root, measure, and record the distances between the consecutive marks. Pin the seedling in moist air, or, when the ink is dry, replant it in a seedling "observation-box" (Appendix B) in moist coco-fibre or sawdust, with its root vertically downward and so near to the glass face as to be visible from the outside of the box. Measure, at intervals of twelve or twenty-four hours, the amount of elongation of the several marked zones. Tabulate the results and plot them in the form of a curve. Determine that the elongating region begins about 1 mm. behind the root apex, extends over a very short section (10-20 mm.), and ends near that part of the root where the root-

hairs begin.   Note also that, further back, the root, though
it has ceased to grow in length, increases in thickness.
Measure the distance from the apex of the region of
maximal elongation.

72.* Cut longitudinal sections through the root-tip and
through the region of greatest elongation : note (if necessary staining with iodine) that the cells of the former region
are small, and that the protoplasm occupies the whole of the
cell; that the cells of the latter are large, that the protoplasm forms a layer against the cell-wall, and that it thus
encloses a large space (vacuole) containing liquid (cf. p. 99,
Fig. 20).   Observe that, when treated with iodine, the
protoplasm, which is killed by the reagent, contracts away
from the wall, and, staining yellow or yellow-brown, becomes readily recognisable.   The extreme tip just beneath
the root cap, composed of small cells with much protoplasm, constitutes the formative region, i.e. the region
where the protoplasts are undergoing division, and thus
forming new protoplasts.   Behind the small formative
region (of about 1 mm.) is the region of elongation
(some 20 mm. in length), and in this region, where some
cell-division also occurs, the increase in length of the
whole root is effected by that of the constituent cells.
We can now understand how, from the very start, the
growth of the root makes for the steadiness of the plant
and secures its fixation.   The old part of the root—the part
in connection with the stem—having ceased to elongate, is
wedged, as in a vice, in the soil.   Nor does the tip elongate.
The growth in length of the region behind the tip results
in a pressure which, since the apical region is free to move,
drives the tip through the soil; the shortness of the elongating region serves to prevent that region, in the event of
the tip meeting with considerable resistance, from being
itself bent.   Each day, a section ceases to elongate, and
each day, a newly formed region just behind the growing-
point begins to increase in length.   Somewhat in the way
that a worm, fixed by its tail-end, pushes, by the elongation
of its body, its head from a burrow, a root pushes its tip
through the soil.

73. By the use of a seedling observation-box, on the
glass front of which we trace each day the outline of the

roots, we follow the course of elongation of the main and
lateral roots of a bean or pea seedling.

We learn from these observations that each root consists
of the following regions : a formative region (growing-
point) covered by the root cap ; a region of elongation ; a
region of thickening, the lower part of which constitutes
the root-hair region.

74. In order to compare the mode of elongation of the
shoot with that of the root, mark the young shoot of a
seedling bean or pea, from base to apex, with a series of
india-ink marks about 8 mm. distant from one another.
Proceeding as in the case of the root, determine the
amount and distribution of the growth of the marked shoot.
We thus discover that the growing region of the shoot is
as extensive as that of the root is limited : that, in the
young stem, parts far distant from the apex, as well as
parts near it, are in active elongation : that the parts
which elongate most rapidly are situated between the
nodes, *i.e.* the places of insertion of the leaves, and that
therefore these internodal regions come each to be of
considerable length. The stem in cleaving the thin air
encounters but slight resistance, and is therefore under
no necessity, as is the root, to concentrate its elongating
region.

The significance of the root-cap now becomes evident.
Behind it, lies the delicate growing-point, in front,
are grains of sand, small stones and other hard substances of the soil which the tip is bound to encounter. As
a thimble protects the finger, so the root-cap protects the
growing-point of the root. It can be shown, moreover, by
simple experiment that the root-tip has another and more
subtle means of protecting itself from mechanical injury.

75. Plant bean seeds in a seedling observation-box
containing, beside coco-fibre, a layer of small stones on
the top of which more coco-fibre, in which the seeds are
planted, is laid and pressed down. Follow the behaviour
of the roots when they come to the layer of pebbles. Note
that, when obstructed, the root curves in such a way that
the tip is first moved away from the obstruction, and then,
brought again into the vertical position, resumes its
downward progress. Determine that the curvature in

response to contact and that which results in the resump-
tion of the vertical position, take place, not at the tip
itself, but in the region of elongation. Some of the
plants raised for the purpose of the above experiments
may be allowed to grow on, in order that the origin and
order of development of the lateral roots may be observed.
As already indicated, the lateral roots are identical in structure with the main root, and grow in precisely the same
manner. Thus each lateral root, wedged like the main
root firmly in the soil, contributes to the firm hold which the
whole root-system has on the soil.

76. Onion or hyacinth bulbs, started into growth in
hyacinth glasses, or similar vessels filled with water, provide objects for the study and the development of another
type of root-system. Observe that a number of independent
roots which, since they arise on the bulbous stem (at its
base), are called adventitious roots, make their appearance
and grow all at about the same rate. Examine the root-
systems of several Dicotyledons and Monocotyledons and
refer them to the bean type (tap root-system, cf. Fig. 18),
or to the onion or lily type (adventitious root-system, Fig.
19). Note also cases (*e.g.* sunflower) in which the tap root
soon ceases to grow, its place being taken by numerous
true, lateral roots (fibrous roots). Observe that plants
with a tap root are deep-rooting (mallow, etc.); those
with the adventitious types of root are more surface-
rooting. Note the curious behaviour of the roots of various
rosette plants (dandelion and plantain), which, after penetrating the soil to a certain depth, shorten so much as to
throw the outer tissues of their older parts into transverse
folds, and observe that, as a consequence, the rosette of
leaves and the epicotyl are dragged down close to the
ground. Note, on a lawn, how the plantain and the dandelion, by this root-contraction, press down the grass in their
neighbourhood, and so secure space and light for themselves.

77. Dig up a dandelion plant. Cut its roots into
lengths : plant the pieces of root in moist, sifted soil.
Observe that, after some weeks, each root has developed
into a plant. Follow the process by which this comes about.
Note that the wounded, cut surfaces are made smooth and

healed by the formation of tissue, called callus or wound-tissue, that, later, roots develope from the new tissue of the lower end, and that buds, which are called adventitious buds, develope from the new tissue of the upper end. We conclude, therefore, that the morphological "points" (p. 85), which distinguish roots from shoots, *e.g.* that the root does not bear buds, are the expression of general, but not of irrevocably fixed habits on the part of these members. We note, in confirmation, that many trees, *e.g.* elm, may throw up suckers from injured places on their more superficial roots.

78. In order to study further the process of healing of wounds and the formation of adventitious roots from callus-tissue, cuttings of geranium, willow, rose, etc., should be made, planted, and examined in their different stages of growth : mount a typical series of such cuttings, and add them to the museum. Observe and record the healing of wounds made when large branches are lopped from trees. If possible, obtain a series of photographs of such wounds and of the appearances they present in the slow course of their healing : put the date and other details on the photographs, mount them, and add them to the museum.

Having studied the fixing function of the root-system, we must next direct our attention to an even more fundamental part of its work, that of absorbing substances from the soil. From general knowledge of plants and gardens, such as the wilting and dying of plants in periods of drought, and from the large percentage of water contained in plant-tissues (see Exp. 5), we are bound to infer that a plant absorbs water from the soil. We shall have an opportunity later (Chapter ix.) of determining the quantities of water taken up by the roots, and will now confine ourselves to enquiring how the absorption of water is effected.

When growing hyacinth or tulip bulbs in water in hyacinth glasses for the purpose of Exp. 76, it may have happened that, owing to neglect in keeping the glasses quite full of water, the later-formed roots found themselves compelled to grow some distance through the air before reaching the water. If such was the case, it may have

PLANT PHYSIOLOGY

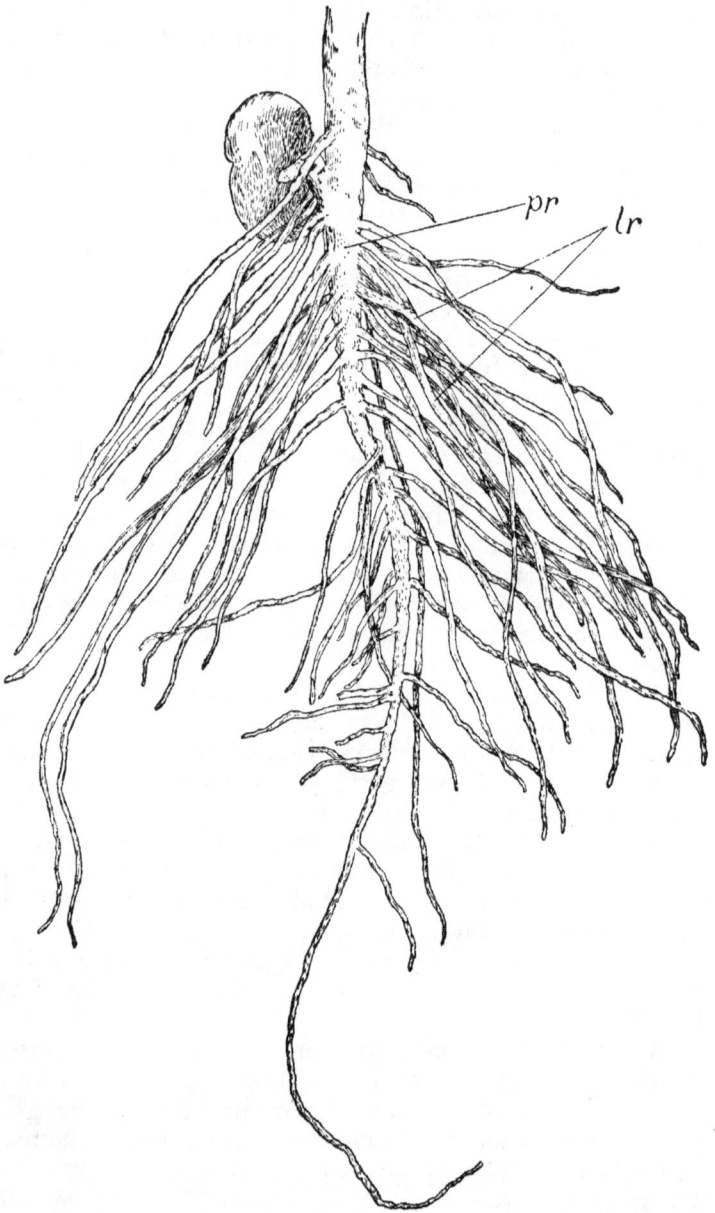

FIG. 18.

Broad bean (Vicia Faba). Root system of seedling, showing elongation of the radicle to form a tap root (*pr*); with numerous lateral roots (*lr*).

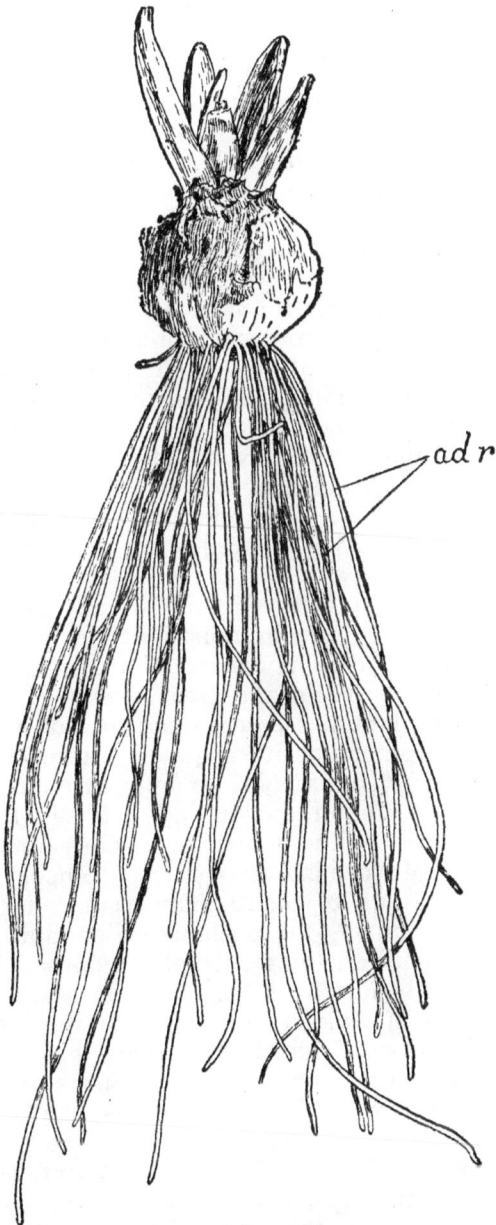

FIG. 19.
Hyacinth (Hyacinthus orientalis). Bulb with adventitious root system;
adr, adventitious roots,

been noticed that, whereas the surfaces of the roots which were submerged from the first are quite smooth, those of the roots which grow through the air are covered with fine, white root-hairs.

In order to obtain material for the study of this phenomenon, we germinate oat or bean seeds, some with their roots in water, others with their roots in moist air.

79.* Make transverse sections across the beginning of the root-hair region of a root growing in air, mount in water, and examine under the microscope. Observe, with low and high powers, the root-hairs in all stages of development. Put the preparations aside and make others through the corresponding regions of a root grown in water which presents to the naked eye no sign of root-hairs. Determine, by microscopic examination of the sections, that slight, dome-shaped outgrowths from the superficial cells occur, and that they are identical with those on the youngest part of the root-hair region of the root grown in moist air.

80. Germinate two lots of maize or oat grains; one, so that the young roots are in moist air; the other, so that they are in moist sand or sawdust. Contrast the development of the root-hairs in the two cases. If, now, we assume that the plant possesses a certain power of regulating the development of its organs in accordance with external conditions, we are led to infer from these facts that the root in water or thoroughly wet soil, being able to carry on its work without their help, does not put itself to the trouble of forming root-hairs. Since, however, countless generations of plants have required root-hairs and have formed them, these plants—like land-plants in general—have acquired the fixed habit of producing root-hairs. Hence a hyacinth or bean root in water is, as it were, the subject of divided counsel. The ingrained habit urges the formation of root-hairs. They begin to form. The presence of much water renders root-hairs unnecessary. They cease to develope. We may suppose that the formation of the incipient root-hairs is due to habit; but that the root, before it obeys the dictates of habit, requires a special stimulus, perhaps, in this case, a slight drying of its surface. In water, this stimulus is lacking, and the root-hairs never get beyond the rudimentary stage. The bodies of plants and of animals contain many rudimentary structures : the calyx of the flowers of the daisy family is often represented by mere hair-like structures : whales and various snakes have hind limbs which are ridiculously small, and are never used : our ears have sets of muscles which do not work : the vermiform appendix, the rudiment of an accessory stomach, is a nuisance. The hypothesis just suggested to account for the rudimentary root-hairs of the water-roots of hyacinths helps us, when applied to these cases, to form some idea why rudimentary organs may outlive their uses. The habit of laying them down belongs to the race : the business of their further development, to the individual. This further development depends on some precise stimulus either from without or from within. In the absence, more or less complete, of this stimulus, the organ remains rudimentary.

Our observations on the presence or absence of hairs on the hyacinth roots lead us to conjecture confidently that the root-hairs are organs which serve to secure the absorption of ample supplies of water. We argue thus : roots grown in water have access to plentiful supplies, roots in moist air to but meagre supplies of water. If the special business of the root-hairs is to secure an adequate amount of water, then, in the case of water-roots, root-hairs will have but little to do ; but in the case of air-roots, their water-absorbing capacities will be exercised to their full extent.

81. We test this hypothesis by an examination of the roots of various water-plants, and discover that, in the large majority, *e.g.* water buttercup (Ranunculus aquatilis, etc.), root-hairs are absent : though, in the case of some aquatic plants, such as the frog-bit (Hydrocharis Morsus-ranae), root-hairs are produced in large numbers. (The aquatic plants collected for this purpose may be examined also with respect to their shoot- and root-systems, which should be compared with those of land-plants, and, with examples of the latter, dried and preserved in the museum, together with notes of their several characters.)

We may take it, then, as probable that the function of root-hairs is to secure a plentiful supply of water for the plant; that, when the root finds itself in water or wet soil, the root-hairs are not required, and may remain rudimentary; but, when the root is in ordinary soil, in which there is no excess of moisture, root-hairs develope and play an important part in the work of water-absorption.

A naked-eye examination of root-hairs suffices to show that they are extremely delicate structures : and microscopic examination confirms this. How delicate they are we demonstrate by pulling up and exposing several of the oat or maize seedlings (of Exp. 80) to the dry air of a room. We observe that the root-hairs soon shrivel and die. Herein lies the explanation of the reason why gardeners, when transplanting seedlings, and indeed any actively-growing plants, are careful to perform the operation quickly, and to leave a ball of earth attached to the root : for, otherwise, the root-hairs are injured and the plant suffers owing to its inability, till a new crop of root-hairs is formed, to obtain adequate supplies of water. We can understand, also, why cuttings, which have, of course, no special absorbent organs, have to be planted in moist soil and protected from loss of water.

82. The maize and oat seedlings of Exp. 80 serve to show that root-hairs are not only delicate, but ephemeral structures. By making sketches to scale of the roots daily, and indicating in the sketches the root-hair region, we discover that, although day after day the extent of the root-hair region remains fairly constant, this is due, not to the permanence of the individual root-hairs, but to the formation of new hairs, just behind the elongating region, to take the place of those which, on the side distant from that region, are shrivelling and dying. Each day, the root-hairs in the older part of the root-hair region wither, and, each day, others are formed in the younger part of that region. If, therefore, we determine the time that the root takes to increase its length by an amount equal to that of the root-hair region, we have determined also the average length of life of a root-hair. By measurements of this kind we discover that the life of a root-hair is generally a matter of a day or two. Since the root of a plant bears root-hairs throughout its life, it follows that, during that period, countless numbers of root-hairs arise, do their work, and die.

The evidence which we have brought together in support of the water-absorbing function of the root-hairs, though strong, is not direct. We have now to attempt to prove definitely that a root-hair cell possesses the power of absorbing water ; and, if we succeed in this, we shall have to enquire how the absorption of water is effected. Our enquiry will necessitate the examination of root-hairs, and also other cellular elements of plants, and will give us information which will be of service to the understanding, not only of the mode of water-absorption by root-hairs, but also of many other phenomena exhibited by plants—as, for example, the method of growth of cells, the origin of the pressure exerted by roots in penetrating rigid soils, the means by which delicate seedlings hold themselves upright, and so on.

83.* We begin this work by making a more thorough examination of preparations of root-hairs (see Exp. 79*). On examining them microscopically, we observe that the first sign of the root-hairs is a slight raising up or projection of the free, outer wall of a surface cell of the root-hair region. We recognise that, since the cellulose wall of a cell is dead, the rounded projection must be due to pressure from within. This pressure continuing, the cell-wall bulges outward till it forms a finger-like projection. By treating preparations with iodine solution, we find that, as in all other protoplasts, that which constitutes the root-hair cell consists of a mass of protoplasm. In the very young cell, the protoplasm forms a more or less dense mass, taking up practically the whole of the space occupied by the cell; but, in the older root-hair cell, the protoplasm, unrecognisable, perhaps, before the addition of the iodine solution, is seen, after its action, to form, and, indeed, to be confined to, a thin layer just beneath the cellulose wall. Within this protoplasmic layer is a clear space or vacuole occupied by a watery fluid, the cell-sap (see Fig. 20). As the action of the iodine continues, this layer, or film of protoplasm stains brown and contracts away from the wall.

84.* In the next place, in order to satisfy ourselves that structures like root-hair cells are able to absorb water, we pull off some of the violet hairs which occur on the filaments of the stamens of Tradescantia virginiana, or, failing this material, we make sections parallel to the surface of the leaves of Primula sinensis (or cross sections of the red stem of this plant) : surface sections of the petals of coloured flowers : or, if nothing better is obtainable, we use thin sections of beetroot.　Our object in choosing such objects is to obtain cells which are similar in structure to root-hair cells, but in which the cell-sap is coloured, and hence readily recognisable.　Examine microscopically and draw the preparation : observe the cell-wall and vacuole with its coloured sap : and note that no protoplasm is to be seen.　Now run in under the cover-glass a 5-10 % solution of common salt, or a 3-5 % solution of potassium nitrate (Appendix A).　Note that, as the salt solution passes into the cell, the coloured sap of the vacuole contracts away from the wall and comes to occupy a smaller space than before.　Since the vacuole contains fluid, and since fluids are incompressible, it follows that the addition of the salt solution has caused some of the liquid of the cell-sap to pass out from the vacuole. But the space which now exists between the cell-wall and the coloured blob of cell-sap is colourless, a fact which indicates that none of the dissolved pigment to which the sap owes its colour has escaped from the vacuole.　By comparing the appearance of the cell in water with that of the cell acted on by salt solutions of increasing strength (Fig. 20), we are led to the conclusions (1) that the vacuole is surrounded by a membrane through which water may pass, but through which the dissolved pigment does not pass.　(2) That, in the intact cell, the membrane is pressed close against the cellulose wall, in which position, owing to its thinness and transparency, it may be invisible.　We treat the sections in salt solution with iodine solution, and observe that the layer in question gives the yellow-brown reaction characteristic of protoplasm. As the action of the iodine continues, observe that the protoplasmic layer, now killed, shrinks away from the cell-wall, and, becoming disorganised, allows the contents of the vacuole to escape.　We may note, also, the deeply-stained nucleus lying either in the protoplasmic membrane or slung in bridles of protoplasm stretching across the cell from one part of the membrane to another.

85.* Make another microscope preparation of the hairs of Tradescantia, etc., run in dilute salt solution, and, whilst the coloured sap is still enclosed in the protoplasmic layer of the cell, add a drop or two of water to one edge of the cover glass, and, by means of blotting-paper placed at the other edge, withdraw the salt solution so that the cell is

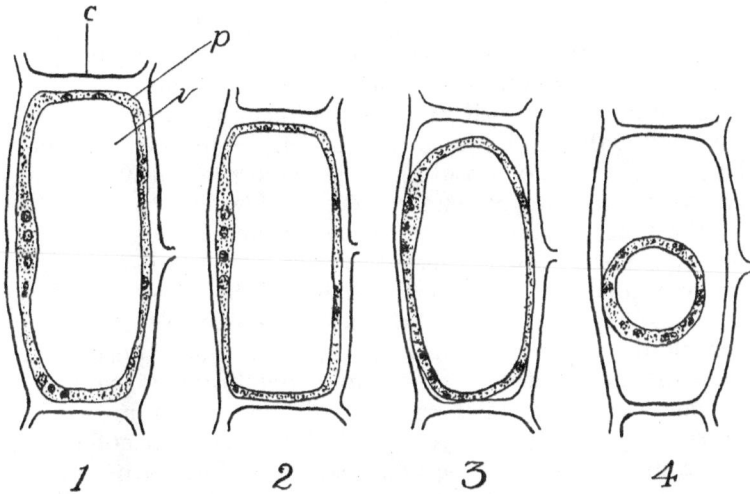

FIG. 20.—YOUNG PLANT CELLS FROM THE TISSUE OF A FLOWER PEDUNCLE. TO SHOW STAGES IN PLASMOLYSIS.

1. In water.   2. In 4 per cent. solution of potassium nitrate.   3. In 6 per cent. solution of potassium nitrate.   4. In 10 per cent. solution of potassium nitrate.   c, cell wall; p, protoplasm; v, cell vacuole.

(After De Vries.)

again in water.   Note that the vacuole increases in size till once again its outline comes to lie close against the cell-wall.   We thus learn that, when the salt solution is replaced by water, the latter substance passes through the membrane of protoplasm (or plasmatic membrane as it is called) to the vacuole.   Thus, the volume of the cell-sap increases and, as a consequence, the membrane is stretched till it comes to lie close against the cellulose framework of the cell.   We infer that, just as salt solution withdraws water from the vacuole, so something in the vacuole may cause water to pass into it from the outside.

It follows from these facts that a vegetable cell is an apparatus of a remarkable nature, and that it consists of a protoplasmic layer across which water may pass, but across which other substances, *e.g.* dissolved pigment of the cell-sap contained in the vacuole, may be unable to pass. This layer is stretchable. Hence if water passes into the vacuole, the increased pressure thus set up stretches the protoplasmic membrane till it is pushed firmly against the cellulose wall. Since also the cellulose cell-wall is stretchable—though less so than the protoplasmic membrane—if more water passes into the vacuole, the cell-wall itself is stretched and the cell becomes larger (cf. Fig. 20, 1 and 2). If, whilst the cell-wall is thus stretched, the protoplasm secretes a new layer of cellulose on the inner surface of the old layer, and if this new layer is sufficiently rigid to maintain the whole cell-wall in its present position without the help of the pressure of the cell-sap in the vacuole, then this distension of the cell by the fluid pressure of the vacuole is made permanent; and so, if now the protoplasmic membrane is caused to shrink away from the wall, the latter, being more rigid by reason of the new layer of cellulose, does not collapse as much as before. There has been, therefore, a permanent increase in size of the cell: in other words, the cell has grown. In the case of the root-hair cell, the stretching of the wall takes effect only at the dome-shaped projection which, in consequence, comes to form the tubular outgrowth characteristic of the root-hair.

A living cell, the vacuole of which is so full of sap as to press the plasmatic membrane close against the cellulose wall of the cell and to stretch the latter to its full extent, is said to be turgid, and is described as being in a state of turgidity or turgor. If water is withdrawn from the vacuole, as, for example, by exposing the cell to the action of salt solution, the volume of the cell-sap decreases and the vacuole becomes smaller. The previously stretched protoplasmic membrane recoils, as a piece of stretched elastic recoils when released, and so contracts away from the cellulose wall. A cell in this condition is said to be plasmolysed, and the process by which this condition is reached is called plasmolysis. The turgid cell may be compared with an inflated bicycle tyre, and the plasmolysed cell with a deflated tyre. As in the inflated tyre, the pressure of the air in the inner tube distends its wall, presses it against that of the outer tube, and stretches the wall of that tube so that the tyre becomes rigid; so, in the turgid cell, the pressure of the fluid in the vacuole stretches its wall (the plasmatic membrane), and, pressing it against the cell-wall, stretches that wall so that the cell becomes rigid : and as a reduction of air-pressure in the inner tube—due to escape of air—results in the recoil of the inner tube, and a relaxation of the distended outer tube, with the result that the tyre becomes flabby, so a reduction of fluid-pressure in the vacuole—due to loss of water— results in the contraction of the previously stretched plas- matic membrane, and consequently in a relaxation of the cell-wall, with the result that the cell loses its rigidity. Consequently, if the elements of a cellular tissue are turgid, the tissue has a certain rigidity; if they are plasmolysed, the tissue becomes flabby.

Since, for reasons which will appear presently, these matters of turgidity and of plasmolysis are of fundamental importance, we will determine by experiments on various plants that the facts we have learned from the study of cells with coloured sap hold good for living vegetable cells in general. We note, for example, the supple sturdiness of delicate seedlings.

86. Lay one or two cress seedlings on the table, and observe that, as they wither, they become flabby. As they dry, the cells lose water and the protoplasmic membranes cease to press on the cellulose walls. The walls are too delicate of themselves to form a sufficient scaffolding or skeleton for the plant, and the seedling is too young to have formed any supporting internal skeleton of woody tissue. Hence the whole plant becomes soft and droops. Note the wilting of leaves of plants in dry weather and infer the cause.

87. Take two bean seedlings, each with roots about $2\frac{1}{2}$ inches long; make on each an india-ink mark about 2 inches from the tip : record the exact distance between mark and tip.   Plunge one root into a 3 % and the other into a 10 % solution of potassium nitrate; leave for several hours : note whether the roots have become flabby.   Re-measure the distance between the tip and ink mark in each case.   Refer the contraction to plasmolysis.   Plunge the roots into ordinary water.   After some hours, remove and measure them.   The root which was partially plasmolysed in the 3 % solution has recovered its normal length.   It has become turgid.   That subjected to 10 % has not re-covered.   Plant both seedlings and determine that the root of the former grows, but that of the latter does not, it having been killed by the stronger salt solution.

88.   Make an india-ink mark beyond the growing region of each of two bean roots growing in moist air.   Determine the rates of elongation of the roots.   Plunge one root into 4 % solution of potassium nitrate, the other into water. Determine that the root in water continues, and that in the salt solution ceases, to grow in length.   Transpose the roots and observe that the partial plasmolysis due to the salt solution is sufficient to check growth : thus proving that turgor is a condition for growth.

89.   Cut two square chunks of beetroot of about equal sizes.   Wipe their surfaces.   Plunge one into boiling water, the other into cold water.   Note that the sap escapes from the former and not from the latter : the escape being due to the destruction of the protoplasmic membrane. Hence determine the approximate temperature at which the cells of the beet are killed.

90.   Cut a cucumber in slices, throw some of the slices on a plate, and cover them with salt : after some hours, compare their flabbiness with the natural firmness of un-treated slices.   Consider the influence of thus plasmolysing a cucumber on its digestibility.

91.   Select several flower-stalks of dandelions bearing open buds ; remove the flower-head and split a stalk length-wise into four sections : note that the pieces curl so that the inner tissue is outside.   Cut the curls into rings, throw several into water, and note that the rings coil up in the same sense as before.   The coiling in water is due to the fact that the cells of the originally inner surface absorb water, and so increase in volume; whereas those of the

originally outer surface of the stalk absorb no water, or
little, do not stretch, and hence are obliged, by the stretching of the inner rows of cells, to bend. Make a model of
wire and indiarubber tubing to illustrate this bending.

Inasmuch as a tissue, when turgid, is longer, owing to
the stretching of the cell-walls, than when it is plasmolysed,
it follows that we can measure the amount of pressure
necessary to produce this difference of length by first
plasmolysing a piece of a plant consisting mainly of cellular
tissue and then determining how big a pull is necessary
to stretch it to its original length.

92. A cowslip flower-stalk serves well for this purpose.
Two india-ink marks are made at about two inches distance
from one another, the distance between the marks is
measured accurately, and the stalk is put into a strong solution of potassium nitrate (15 %). When the stem has become flabby, the amount of contraction is determined by
remeasuring the distance between the marks. The stalk is
then laid horizontally on a plate of cork or wood; one end,
just beyond an ink mark, is fixed securely to the plate by
means of pins or thin nails, and to the other end, just
beyond the other mark, a stout thread is tied, the thread is
passed over a simple pulley fixed to the edge of the table and
attached to a scale pan taken from a chemical balance.
By placing a millimetre scale beside the stalk, and by adding
weights to the scale pan, determine the weight required
to stretch the stalk to its original length. To express
the result more definitely, measure, *e.g.* by means of a
microscope provided with a micrometer (Appendix A),
the area of a cross-section of the stalk at the beginning
of the experiment, and, neglecting the fact that the stalk
is made up of a vast number of cells, consider it as a
single cell and argue thus :—the pull on the end walls necessary to stretch the plasmolysed stalk to its original length
is so much : but this original length of the turgid stalk was
maintained by the pressure of the cell-sap in the vacuole :
therefore, the latter pressure, causing turgor, is equal to
the mechanical pull due to the weights on the scale pan :
since we know the area of the cross-section we can express
this pressure in pounds per square inch : and since we
know that the atmosphere exerts a pressure of about 22 lbs.

to the square inch, we can express the turgor-pressure in terms of atmospheric pressure, *e.g.* as equal to 1, 2, 3, 4, 5, or more atmospheres. By means of measurements made in this and other ways, it has been found that the turgor-pressure in plant-tissues may be very great indeed, reaching as much as 10, or, in some cases, even 20 atmospheres. It is by reason of these great pressures that relatively delicate plant-structures may, when they encounter obstacles to their growth, perform remarkable feats of strength : thus they may cleave rocks or, like the thistles of the American prairies, raise up and bend railway lines.

Our investigation of the mode of absorption of water by the root-hair cell has carried us a long way from our immediate subject; nevertheless, it has not been irrelevant. For it is evident that, just as the partially plasmolysed cell of Exp. 85 * absorbs water and recovers its turgidity, so may a root-hair cell absorb water and become distended. If a cell neighbouring on a root-hair cell has been losing water, it may withdraw water from the vacuole of the root-hair cell just as the salt solution withdraws water from the turgid cell in Exp. 84 *. Then the root-hair cell, having lost water to its neighbours, may take up from the soil more water to make good that loss. In this way is obtained the large quantity of water absorbed by the root, and in this way also the water, thus obtained, passes from the root-hair cells into the wood-vessels of the root (Bibliography, 5, 6).

CHAPTER VII.

OUR experiments in the last chapter have proved that the cells of plants are capable of taking water into their vacuoles; but they have not shown us how this absorption of water is effected. To the investigation of the mode of absorption we must now address ourselves. Regarding, for the time being, a vegetable cell as an apparatus for the absorption of water, we have to ask how this apparatus works.

The answer to this question will be forthcoming if we can determine by virtue of what property the cell-sap, enclosed within the plasmatic membrane, is able to absorb water.

Now, inasmuch as the same property is exhibited by solutions of salts, sugar, etc. (cf. Exp. 85 *), it may well be that similar solutions occur within the vacuoles of plant-cells. Indeed, we can demonstrate, by employing the copper sulphate test, that the cell-sap of the cells of beetroot contains sugar, and, by other appropriate tests, that salts, both organic and inorganic, occur in the cell-sap of the cells of plants. Thus our question becomes : by virtue of what property may such substances as organic and inorganic salts, sugars, etc., cause water to pass across a membrane like that which occurs in the living cell?

Everyone is familiar with the phenomenon of diffusion of gases; or, at all events, with manifestations of this phenomenon : for example, the pervasion of a whole room by the odour of Eau de Cologne or musk or other scent when the stopper is taken out of the bottle in which it is contained. The rapid dissemination of the odour is doubtless due to air currents which carry the odoriferous substance : but, even if the air of the room was perfectly still, the vapour given off by the scent would be distributed gradually throughout the space.

In order to account for the phenomenon of gaseous diffusion, and also for other phenomena exhibited by gases, a theory—the kinetic theory of gases—has been put forward, which suggests that, though, in any gas, there are no visible mass movements, nevertheless the invisible molecules are in a state of constant movement. Applying this theory to liquids, we may suppose that, in a solution of salt, the molecules of salt are moving through the water in all directions like those of a gas liberated in the air, though, owing to some form of obstruction on the part of the water molecules, not so freely. From this hypothesis it should follow that if a solution of salt is brought into contact with water without shaking, the salt molecules travel gradually throughout the water till, ultimately, they become uniformly distributed in it. That this actually takes place, we demonstrate in the following way :—

93. A long cylindrical jar is filled with water and stood on a steady table. By means of a funnel (*e.g.* thistle funnel, see Appendix A) a strong solution of a coloured salt, *e.g.* bichromate of potassium is poured into the vessel so as to form a distinct layer at the bottom. A white paper millimetre scale is fixed behind the vessel and a daily record taken of the height reached by the potassium bichromate.

This way of measuring the rate of diffusion is open to the objections that changes of temperature set up convection currents in the liquid, and that, unless the table on which the vessel stands is perfectly steady, vibration, caused by people walking in the room, etc., increases the rate at which the liquids mix. Both sources of error lead to a fictitiously high result.

The following method is simpler to carry out and is not open to these objections :—

94. To leaf gelatine, obtainable in commerce, add enough water to form a liquid which, when it cools, sets to a firm jelly. Before the liquid gelatine cools, add a small quantity of thymol or other antiseptic (Appendix A), and also a few drops of phenolphthaleïn (Appendix A), a substance which undergoes a marked colour change— from colourless to rose—when acted on by an alkali. Pour the warm, liquid gelatine carefully into a tall jar and, when it has set, invert the jar over a dish containing a fairly strong solution of potash (KOH) or soda

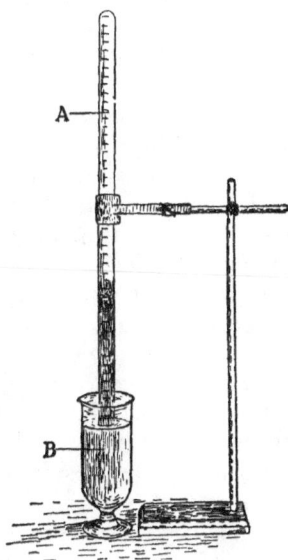

FIG. 21.—APPARATUS TO ILLUSTRATE DIFFUSION OF LIQUIDS.

A graduated glass tube (A) containing gelatine and an indicator (phenolphthaleïn) is inverted in a vessel (B) containing a solution of potash.

(NaOH), which give a strong alkaline reaction. Determine the rate of diffusion of the alkali by tracing the ascent of the rose colour. Record the results. The fact that the experiment lasts for weeks should not lead to neglect in the matter of recording; for it is important for us to learn that this liquid diffusion is a *slow* process. The experiment demonstrates also that the gelatine does not prevent the diffusion of water and potash.

We know from previous experiments (*e.g.* Exp. 26) that a similar intermixing of water and sugar takes

place when these substances are separated from one
another by a parchment membrane. Nor is this surprising if we suppose that diffusion of gases or of
liquids is due to molecular- and not to mass-movements.
Although such a parchment membrane, when used to contain water, does not leak, its substance must be pictured as
riddled with a system of ultramicroscopic spaces, too small
to allow the smallest drop of water to pass, but amply
large enough to allow of free movement of molecules of
H_2O. Inasmuch as the molecules of fluids are, according
to our assumption, like those of gases, in constant movement, some water-molecules pass into the interstices of
the membrane, and, arriving at the other side, come
in chemical contact with the sugar-molecules. The fact
that sugar is soluble in water means that sugar- and water-
molecules exercise some attraction on one another. We
may conceive of the sugar-molecule linking with the water-
molecule and joining it in the molecular dance; though,
being heavy, the sugar-molecule will cause a slowing of
the rate of movement. The molecular dance is erratic.
Sugar and water partners career in all directions, some
back through the interstices of the parchment membrane,
some on through the sugar solution. Since the molecular
movements are unending, the result is that, in course
of time, there are as many sugar-molecules per unit of
volume on one side of the membrane as on the other.
Looking at the final result, we say that sugar and water
have passed by osmosis each from one side of the membrane to the other. Even after equilibrium has been
reached, that is, when the liquid on one side of the membrane contains per unit volume as much sugar as that on
the other side, the molecular movements do not cease.
But, in these circumstances, the number of sugar-molecules
conveyed in a given time across the membrane in one
direction is equal to the number which pass in the opposite
direction. Just so may the population of a town remain
remarkably constant though the townsmen come and go.

In order to study the phenomenon of osmosis more
closely, with the object of learning more about the mode
of absorption by the plant, we make an osmotic apparatus
thus :—

95. Cut off a length of about eight inches of parchment tube (Appendix B). If dry, soak the parchment tube in water : when thoroughly wet, fix in one end a rubber cork without holes and make the joint good by winding tightly round the parchment tube many turns of stout silk thread or florist's buttonhole wire. Stand the tube upright, open end uppermost, in a pail or other suitable large vessel. Make two holes in opposite sides of the parchment on a level with the rim of the pail. Pass a glass rod through the holes, and let the rod rest on the rim (Fig. 11). Fill the outer vessel with water up to about two inches below the level of the glass rod, then fill the parchment tube nearly up to the level of the holes through which the glass rod passes. Leave the apparatus for a short time and observe whether any water has escaped from the tube. If so, there is a hole in it, and the tube must be replaced by another. If it proves to be sound, empty the tube and cut away the part pierced by the holes through which the glass rod passed. Choose a two-holed rubber cork which just fits the tube. Insert through one hole the tubular limb of a separating funnel, seeing that it makes a good joint. Through the second hole in the cork, pass a glass tube about three feet long so that its lower end is flush with the lower surface of the cork. Stand the empty parchment tube in the vessel of water, insert the cork, and make good the junction between cork and tube by many turns of silk thread, supplemented, if necessary, by wax (Appendix A). Fix the glass tubes, or the longer one only, in a clamp, so that the lower cork rests on the bottom of the outer vessel and the parchment tube is steadied (Fig. 22). Prepare a strong solution of sugar : colour the sugar with a dye such as methylene blue. Pour the coloured sugar solution into the separating funnel. Open the stop-cock of the latter and run in the sugar solution till, the parchment tube being full, the liquid rises in the long glass tube to about half an inch above the level of the cork. Close the stop-cock of the separating funnel. Pour water into the outer vessel to about the level of the top of the parchment tube. If any sugar solution has been spilled into the outer vessel, transfer the osmotic apparatus bodily to another, similar vessel of clean water.

The apparatus is laborious to make, but is so instructive
in action that no trouble should be spared to get it set up.
An improvement which enables us to use it again and again
is illustrated in Fig. 22. The outer vessel, which serves

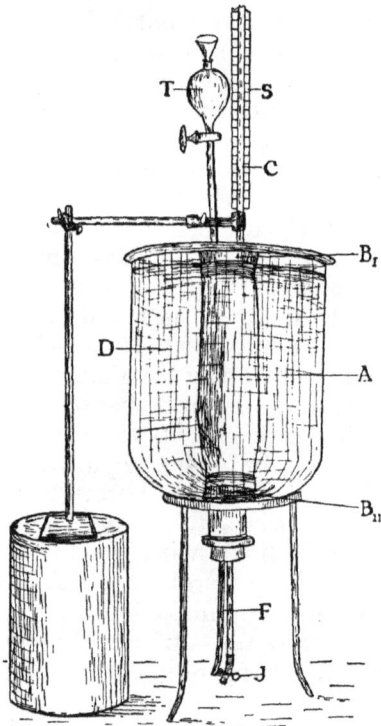

FIG. 22.—APPARATUS TO DEMONSTRATE OSMOTIC PRESSURE.

A, parchment tube; B_1, B_{11}, rubber corks; T, thistle funnel; C, glass tubing;
D, vessel containing water; F, exit tube; J, clamp; S, scale.

to contain the parchment tube, consists of an inverted
bell-jar with an open neck (Fig. 22 D), and is supported
on a strong iron tripod-stand. The lower end of the
parchment tube is closed by a rubber cork (B_{11}) with one
hole, into which is fitted a short glass tube, the upper end
of which is flush with the upper surface of the cork. The
tube is passed also through the hole in a rubber cork which

closes the neck of the bell-jar. To the free end of the short glass tube (F), a short piece of rubber tubing is attached and closed by means of a clamp (J). When the upper cork (B_1) with its funnel and long glass tube has been inserted in the upper end of the parchment tube, the bell-jar is filled with water and the apparatus is ready for use. The coloured sugar solution, to which a trace of an antiseptic may be added, is poured through the funnel and the air contained in the short tube F is expelled by opening the clamp J. After the apparatus has been used in the way about to be described, the liquid in the parchment tube may be withdrawn through the tube F and the parchment tube refilled with sugar or other solution, the osmotic properties of which it is required to test.

An osmotic apparatus of this kind, once fitted up, will last for months if the parchment tube is not allowed to become dry, and may be used for various interesting experiments in osmosis.

If all the joints are good and the parchment tube intact, the fluid begins to rise almost at once in the long glass tube. Fix a paper scale behind the long tube and take time-records of the height to which the liquid rises. After some time, it becomes necessary to add another length of glass tubing, which is connected with the long tube by a short piece of rubber tubing and supported, e.g. against the wall of the room. The liquid continues to rise to a height of many feet. Then, after some time, it begins to fall and finally descends to the level of the liquid in the outer vessel. Before this takes place, note that the dye has escaped through the parchment-wall into the water outside, and prove by the sugar test (Exp. 21) that sugar also has passed into the outer vessel.

The rise of the water in the long glass tube means that the volume of liquid in the osmotic apparatus has increased, and hence that water has passed through the parchment-wall. Therefore, in the course of osmosis there is, as indeed we should expect from our previous studies, a movement in both directions : sugar and soluble pigment pass from parchment tube to the outer vessel; water passes into the parchment tube. Since, moreover, the liquid rises in the long tube, the amount of water which passes in is greater than the amount of sugar which passes out.   We may say that the sugar solution exercises an osmotic attraction for water, and that it sets up an osmotic pressure sufficient to hold up the high column of water in the long tube.

But since, from the moment the apparatus is set to work, sugar begins to diffuse out into the surrounding liquid, and since, in that situation, it sets up a counter-attraction, tending to withdraw water from the parchment tube, our apparatus is useless as an osmometer,—that is, as a measurer of the actual osmotic pressure exerted by the original sugar solution.   It serves only to indicate that such pressure exists and can be made to do work, e.g. of lifting water.   That the sugar which escapes into the outer vessel does set up a pressure tending to neutralise that in the parchment tube we demonstrate thus :—When the liquid in the long tube has reached a certain height and has become fairly stationary, siphon off that in the outer vessel and replace it by water.   Note that the liquid in the long tube begins again to ascend.   The level previously reached represented the resultant of the osmotic pressures exerted by the sugar solutions inside and outside the parchment tube : by removing the sugar from the outer vessel, the pressure exerted by that in the tube is no longer in part counteracted, and so is able to hold up a longer column of water.

96. By using solutions, e.g. of cane-sugar and of grape-sugar successively, we may demonstrate by means of our osmotic apparatus that the rates at which substances in solution pass through a parchment membrane depend, among other things, on their molecular weights. The molecular weight of cane-sugar ($C_{12}H_{22}O_{11}$) is 342; that of glucose ($C_6H_{12}O_6$) is 180; therefore, if we dissolve the molecular weights of cane-sugar and of glucose in grams, i.e. 342 grams of cane-sugar and 180 grams of glucose in equal volumes of water, say 1000 c.c., and use first the one and then the other solution in the parchment tube, we find that the height of the water column supported by the cane-sugar solution is greater than that supported by the grape-sugar.   For the heavier cane-sugar molecules diffuse more slowly than the lighter molecules of glucose.   Hence, before the latter has had time to get up its full osmotic pressure, much of it has diffused into the outer vessel. Similarly if we fill the osmotic apparatus with a solution of potassium nitrate of corresponding strength, *i.e.* one containing the molecular weight of KNO_3 in grams, viz. 101 grams, per 1000 c.c. of water, we find that, owing to the high rate of diffusion of potassium nitrate, the osmotic pressure which the solution sets up is very small indeed and by no means a true measure of that which it can exert.

We have, however, in plant-cells and tissues an apparatus ready-made for comparing the osmotic pressures exerted by different substances.   For a vegetable cell is an osmotic apparatus strikingly similar to that used in Exp. 95.   The osmotic substances dissolved in the cell-sap correspond to the sugar or other solution in that apparatus, and the proto-plasmic membrane (plasmatic membrane) of the cell corre-sponds to the parchment membrane.   If a completely turgid vegetable cell is placed in water, it undergoes no change in volume since the cell-wall is already stretched to its limit by the pressure of the cell-sap : if, however, the cell is not completely turgid, water is absorbed, and the volume of the cell is increased—the osmotic pressure being used to do the work of stretching the plasmatic membrane and cell-wall.   If such a cell is placed in a strong salt solution, the osmotic pressure of the latter causes water to pass out from the vacuole, across the protoplast and cell-wall.   The amount of fluid in the vacuole being thus reduced, the pressure on protoplast and cell-wall falls off and the volume of the cell decreases.   As water is withdrawn, the solution of osmotic substances in the cell-sap becomes more con-centrated, and consequently the osmotic pressure of the cell-sap increases.   If the osmotic pressure of the salt solution used is greater than that of the concentrated sap, more water is withdrawn from the vacuole, plasmolysis sets in, and, yet more water being withdrawn, the proto-plast, shrunk away from the wall, may collapse and become disorganised.   If, then, we place vegetable cells or tissues in salt solutions of different strengths, we find that, whereas, in strong solutions, plasmolysis is complete, and in very weak solutions it does not occur at all, there is one strength of solution which just suffices to produce the early stage of plasmolysis and no more

K.P.   H

(cf. Fig. 20, 2). Evidently this means that the osmotic pressure of the cell-sap is, at this stage, just equal to that of the salt solution outside the cell.

97.* Make up 100 c.c. of a gram-molecule solution (g.m. solution) of potassium nitrate, *i.e.* one which contains the molecular weight in grams of KNO_3 in 1000 c.c. of water (molecular weight of $KNO_3 = 39 + 14 + (16 \times 3) = 101$) : hence the g.m. solution is made by dissolving 10·1 g. in 100 c.c. of distilled water. Note that, the molecular weight of KNO_3 being 101, it so happens that a g.m. solution of KNO_3 is very nearly a 10 % solution. Dilute 50 c.c. of the original solution with an equal volume of distilled water, thus obtaining a $\frac{1}{2}$ g.m. solution (approximately 5 %). Similarly prepare a $\frac{1}{4}$ g.m. solution (2·5 %), a $\frac{1}{8}$ g.m. solution (1·25 %), and a $\frac{1}{16}$ g.m. solution ('62 %). Determine by trial of these solutions on microscope preparations of cells with coloured sap (see Exp. 84*), which of them cause plasmolysis and which do not : then prepare from the above solutions one of known strength which just sets up the first stage of plasmolysis. Record the results, and indicate in the record the strength in gram-molecules of the just-plas-molysing solution.

Make up 50 c.c. of a gram-molecule solution of cane-sugar and determine what strength in gram-molecules just plasmolyses the cells of tissues similar to those used in the experiment with potassium nitrate. Since things which are equal to the same thing are equal to one another, the osmotic pressure of the potassium nitrate solution which just plasmolyses the cells is equal to that of the sugar solution which produces a like effect.

Find the strength in gram-molecules of glucose, $C_6H_{12}O_6$, which is sufficient to initiate plasmolysis in cells similar to those used in the experiment with cane-sugar. Infer from the results with cane- and grape-sugar that the osmotic pressure of cane-sugar is to that of glucose as the molecular weight of the former is to the molecular weight of the latter. Observe that potassium nitrate, when compared with sugar, gives an osmotic pressure about $1\frac{1}{2}$ times greater than we should expect from its molecular weight. If we imagine that KNO_3 in solution exists in part as molecules of KNO_3 and in part of bodies (called ions) consisting of K and

NO_3, and if each of these exercises osmotic pressure, we can form an idea of the reason why the osmotic pressure exerted by a solution of KNO_3 of known strength is greater than we should have expected on the basis of the " law," which we have established for cane-sugar and glucose, that equi-molecular solutions of osmotic substances exert equal osmotic pressures.   It was by means of experiments such as these that the discovery was made that the osmotic pressure of a substance is a property of its molecules in the same way that weight, etc., are properties of the molecules.   These discoveries have led to others of equal interest, and those in turn have served as the basis for various theories to account for the known facts.   We cannot, however, pursue this part of our subject further, but refer students who wish to learn more about the physical aspect of osmotic pressure to the larger text-books (Bibliography, 11).

The comparison of a plant-cell with the osmotic apparatus of parchment tube (see p. 113) fails in one very important particular.   For, whereas such substances as soluble pigments pass readily across the parchment-wall (see Exp. 84*), the soluble pigment in the cell-sap remains, in the partially plasmolysed cell, enclosed within the plasmatic membrane.   There must, therefore, be an important difference between parchment and protoplasmic membranes.   The parchment membrane is indiscriminately permeable to diffusible substances, such as water, sugar, potassium nitrate, soluble pigments like methylene blue, etc., but the protoplasmic membrane is not.   The former kind of membrane is called a permeable membrane : the latter kind, which is permeable to water but not to all diffusible substances, is called a semi-permeable membrane. The importance to the plant of this property of semi-permeability is very great indeed : for, were the osmotic membrane of the plant-cell permeable, there would be a constant osmotic leakage of osmotic substances from the cells of the plant into the soil just as there is a constant osmotic leakage of sugar from the osmotic apparatus of Exp. 95.

98. Instead of using cells with coloured cell-sap, the curved pieces of split dandelion stalk (Exp. 91) may be

employed in the above experiment.   To use them, proceed
as follows :—

Split a dandelion stalk longitudinally into four pieces,
cut the pieces into lengths of about an inch each, and,
having prepared the series of solutions the osmotic pressures of which are to be determined, trace on paper, by
means of a brush and india-ink, the amount of curvature
of each piece of stalk.   Throw a piece into each of the
solutions; after some minutes, take the pieces out, determine which have curved more, and which have become
straighter.

The solutions which have caused a straightening have
done so by withdrawing water from the cells of the dandelion.   Their osmotic pressures are, therefore, greater than
the osmotic pressure of the cell-sap of the cells of the stalk.

Those solutions in which the pieces have curved more
have yielded water to the cells, that is, the osmotic pressure
of these solutions is lower than that of the cell-sap of the
dandelion cells.   Hence, by finding a solution in which
pieces of dandelion stalk retain their curvature unchanged,
we find what strength of solution is in osmotic balance
with the cell-sap.

99.   By using pieces of dandelion stalk in the same way,
we may compare known strengths of different osmotic substances with one another, with respect to their osmotic
pressure.

It is important to demonstrate that the property of
semi-permeability referred to on p. 115 is not confined to
living membranes.

100.   Thus we may make a semi-permeable membrane,
called, because of its mode of preparation, a precipitation
membrane, by causing solutions of copper sulphate and
potassium ferrocyanide to interact.   The precipitate of
copper ferrocyanide which forms as the result of this interaction, is not granular, but skin-like.   To prepare it, pour
a 3% solution of potassium ferrocyanide into a wide-
mouthed bottle fitted loosely with a cork.   Fix a glass
tube, drawn out at one end to a fine point, in a hole in the
cork so that its fine end dips below the ferrocyanide solution.
Having withdrawn and cleaned the tube, draw a drop of
strong copper sulphate solution into it; close the tube

by the finger, and fit the cork attached to it in the neck of the bottle so that the fine end of the tube is just below the surface of the potassium ferrocyanide solution. When the finger is removed from the end of the tube, the copper sulphate reacts with the ferrocyanide to produce a membrane, which closes the fine end of the tube. Inasmuch as the copper sulphate on one side is a strong solution and has an osmotic pressure higher than that of the weak potassium ferrocyanide solution on the other side of the precipitation membrane, water passes from the latter solution to the former. Hence the volume of the copper sulphate solution increases, its increased weight stretches the precipitation membrane and the "artificial cell" grows. As the pressure of the copper sulphate solution increases further, the membrane is ruptured; but it forms again owing to a new skin of copper ferrocyanide being produced as soon as the copper sulphate and potassium ferrocyanide come in contact. Again the "cell" grows, and again it is ruptured to be once more repaired. This membrane, permeable to water but not to such substances as copper sulphate or potassium ferrocyanide, is a semi-permeable membrane.

Another method of making a semi-permeable precipitation membrane is as follows :—

101. To ordinary gum, add a small quantity of gelatine, some sugar and a little colouring matter, *e.g.* aniline blue in solution. Take up a little of the gum-mixture on the rounded end of a piece of thick glass rod and expose it to the air till it is dry. Put the rod so that its gummed end is in a 2 % solution of tannin. A membranous precipitate of tannate of gelatine is formed on the surface of the dried gum. The sugar contained in the gum exerts its osmotic pressure, water passes across the membrane; the latter is stretched and behaves like the membrane in the former experiment. The dye, however, does not pass across. Thus the precipitation membrane behaves like the plasmatic membrane of a vegetable cell (cf. Exp. 84*), permitting water to pass across, but preventing the osmosis of the dissolved pigment.

102. Again, we may illustrate the semi-permeability of precipitation membranes by the use of our parchment tubes.

Prepare two tubes, A and B, supported by glass rods as in Exp. 17.   Into each pour a 1 % solution of calcium nitrate : stand A in water and B in a 1 % solution of disodic phosphate.   Add a little methylene blue to the liquid in each parchment tube.   After a day, note that the methylene blue has appeared in the water of the outer vessel of A; but not in that of B.   In B, the calcium nitrate and disodic phosphate interact to form a precipitation membrane of calcium phosphate on the wall of the parchment tube, and this membrane, though permeable to water, is impermeable to a solution of methylene blue.

On the lines of such an apparatus as that of Exp. 100, a perfect, working, osmotic model of a vegetable cell may be constructed.

Thus we reach the end of our enquiry.   We discover that, by means of its root-hair cells, the root absorbs water together with any soluble osmotic substances to which the plasmatic membrane of the root-hair cell is permeable.   What these latter substances are, we must determine by other methods and in another chapter.

CHAPTER VIII.

The substances taken up by the roots of plants. The composition of plant-ash. Water- and sand-cultures. The soil in relation to plant-life. The origin of soils: their physical, chemical, and biological properties.

The fact, established in the last chapter, that root-hairs are capable of absorbing water and any other soluble substances to which their plasmatic membranes are permeable, leads directly to the enquiry :—What are the substances which the roots of plants absorb, and to what use does the plant put the substances which it obtains from the soil?

That water is absorbed by the root is evident without further experiment; for, on the one hand, we have found (Exp. 5) that plants contain large quantities of water, and, on the other, we know that unless water is supplied to their roots, plants wither and die.

Presently we shall have to enquire more closely into the relation of the plant to water, but, in the meantime, we will devote ourselves to discovering what substances, other than water, are absorbed by the root.

The same experiment which taught us that the tissues of plants contain considerable quantities of water, proved also that the dry matter, left after the water is driven off, consists largely of carbon compounds, and that, after these compounds are burned away, a relatively small amount of mineral matter remains behind as ash.

Now, carbon compounds of the kind contained in the plant do not exist in the soil, but inasmuch as such mineral substances as those found in the ash are always present in the earth, we may be fairly confident that the mineral substances of the ash of plants are obtained in one form or another from the soil.

To determine accurately and completely all the constituent elements contained in plant-ash would require a somewhat elaborate chemical analysis. Though such an analysis is beyond us, we can demonstrate readily the presence of some of the more important elements by the following methods :—

103. Fill a large bottle with fine ashes from a bonfire, or from a bundle of dried hay burned for the purpose.

Place some of the ash on a piece of platinum foil (Appendix A), and heat it in a Bunsen flame. Roll up the platinum foil with the remains of the fused ash, place it in a test tube; add a little distilled water : boil. Test the solution so obtained with red litmus paper : a blue colour shows that an alkali is present. (Use the solution for Exp. 105.)

104. Moisten a platinum wire (Appendix A) with hydrochloric acid, dip it in the ash and hold it in the flame. Observe the yellow colour of the flame, due to the presence of *sodium*. Repeat, observing the colour of the flame through a piece of cobalt-blue glass : a violet colour indicates the presence of *potassium*.

105. To the solution obtained in Exp. 103, add an equal volume of dilute nitric acid. Pour the liquid into a clean test tube, add an excess of a solution of ammonium molybdate (Appendix A), boil: a yellow precipitate indicates that *phosphates* are present in the ash.

By appropriate tests, the presence in the ash of other mineral substances—compounds of *calcium, magnesium,* etc.—may be demonstrated.

It is to be noted, however, that these tests give no indication as to the form in which the elements, potassium, magnesium, calcium, etc., exist in the plant, for the compounds present in the intact plant are decomposed in the process of burning.

Since, however, as our tests indicate, potassium, sodium, calcium, phosphorus, etc., are present in combined form in the ash, these elements must have been present in some form or other in the plants which yielded the ash. Whence it follows that they were obtained from the soil.

The question therefore arises, are the mineral substances which plants contain mere accidental accumulations or are they of significance in plant nutrition?

Careful analyses of plants have proved that the number of inorganic substances which may be present in the plant is considerable. By comparison of the results of analysis of various species of plants, it is discovered that, whilst certain elements are present in all plants, others are not. Thus, analyses of plants growing on some soils show traces of copper, zinc and other rarer elements to be present in combination in the ash. Since, however, plants of the same species, grown in other soils, contain no trace of these particular substances, and yet develope properly, we may regard these occasionally occurring, inorganic bodies as unessential.

If we draw up a list of the mineral elements which are universally, or at all events very commonly, present in plants, we find that it contains the following :—Potassium (K), Magnesium (Mg), Calcium (Ca), Iron (Fe), Phosphorus (P), Sulphur (S), Silicon (Si), Chlorine (Cl), and Sodium (Na).

In order to complete our analysis of the elementary composition of plants, we add the elements present in the gases formed when the dry matter of the plant is burned, viz. : Carbon (C), Oxygen (O), Hydrogen (H), Nitrogen (N), and Sulphur (S).

Inasmuch as it does not necessarily follow, because a substance is invariably present in the plant that it is therefore essential, we require some means, other than that of direct chemical analysis, of determining which elements are essential.

It was a consideration of this kind which led physiologists about fifty years ago to make the plant itself serve to demonstrate which of the commonly occurring elements are essential, and which, if any, are of no service to the plant. This they did by growing plants rooted in media of known composition, e.g. pure water or pure quartz sand, to which were added compounds of the mineral elements of our list, viz. salts containing K, Mg, Ca, Fe, N, P, S, Si, Cl, Na. Plants grown under such artificial conditions were found to flourish quite as well as in the soil. Then, by withholding each element in turn, it could be determined whether the development of the plant was affected in consequence. In this manner it was proved that silicon, chlorine and sodium, though generally present in plant-ash, are not essential for the perfect development of flowering plants in general.

Since water- and sand-cultures are among the most instructive experiments in plant physiology, we proceed to carry them out with a view to ascertaining what symptoms are manifested by plants deprived of one or other of the essential, mineral substances. The salts and also the water used for the culture solutions must be pure. For full instructions as to the mode of preparation of the solutions, see Appendix A.

It will be of interest to repeat, as the first of our sand-culture experiments, the original sand-culture made by Boussingault, who introduced the method about 1868.

106. Procure good silver sand, clean it (Appendix B) : place pieces of clean crock (broken flower pots) at the bottom of each of six, small, new pots and fill the pots with the sand : plant one pea seed in each pot (the seeds, before planting, may be plunged whilst dry into hot water for a few minutes, soaked in tepid water for twelve hours, and then sown). Water thoroughly and cover the pots each with a piece of glass. When the seedlings appear, water three of the pots with complete culture solution (Normal solution, Appendix A), and three with a culture solution complete, except for nitrates (Normal, minus nitrates). When planting the seeds in the pots of sand, plant others in pots of ordinary soil to serve for comparison. Record the rate of growth of the seedlings of the three series : when the difference between those with and without nitrates has become marked, draw or photograph the plants (Frontispiece). From the starveling appearance presented by the plants grown in sand lacking nitrates, and from the healthy appearance presented by those in sand containing nitrates, we conclude that a plant cannot live without nitrogen, that the supply of nitrogen which the seedling peas contain in the form of proteins in the aleurone grains of the cotyledons (cf. Chapter III.) is soon exhausted, and that the plant then turns to the soil for further supplies of nitrogen, which supplies the soil may furnish in the forms of nitrates. Beside thus demonstrating that plants derive nitrogen from the soil, we get a striking example of the complexity of the chemical process carried on by plants. For, from our study of the nutrition of seedlings, we have learned that plants build up their protoplasm from complex, organic nitrogen-compounds (*e.g.* proteins), and now we are driven to the conclusion that the raw materials, which provide the nitrogen contained in the proteins of plants, are the relatively simple, inorganic nitrogen salts which are absorbed from the soil. Though physiologists have spent much time in investigating the process of nitrogen-assimilation, *i.e.* the process whereby the nitrogen of the nitrates of the soil is combined with other elements to form amino-compounds and proteins, our knowledge of the subject remains fragmentary (Bibliography, 10, 11).

Beside its historical interest, this sand-culture experiment of Boussingault has yet another interest, and for that reason we shall return to it presently (p. 141).

107. Repeat Exp. 106, using plants with smaller seeds in order to avoid the complications introduced by reserve-materials. Among the various plants which may be used successfully for sand-cultures, poppies (Papaver sp.) or the Californian poppy (Eschscholtzia californica) are, by reason of the readiness with which they grow in sand, among the best.

108. By means of sand-cultures determine the effect of withholding certain of the other elements of our list, viz. :—calcium, magnesium, and phosphorus (see Appendix A). Though the results are less striking than those obtained by withholding nitrogen, yet, in each of these cases, the plant fails to develope properly.

Mount the drawings or photographic records of the sand-culture experiments on cards : add details and preserve them in the museum. Record peculiarities as to leaf-development, colour of leaves, growth of stem, flower- and fruit-formation in each case.

Unless the sand used in the above cultures is pure, and unless special precautions are taken, an attempt to demonstrate the effect of withholding iron will fail ; and this for two reasons. First, the amount of iron required by plants is extremely small, and second, traces of iron occur in most samples of sand. Hence to demonstrate the effect produced in the plant by lack of iron we resort to the method of water-culture (Appendix A).

109. Procure several good-sized glass bottles—large jam bottles answer the purpose well. Clean and dry them and immerse the bottles almost to their rims in hot water of a temperature of about 50° C. Melt a little solid paraffin (of melting point 40° C., see Appendix A) in a porcelain dish, and pour it into one of the bottles. Rotate the bottle so that the paraffin, as it cools, forms a thin layer over the whole of its inner surface. Treat the other bottles in a similar way. Having germinated maize so that the roots of the seedlings are in water, remove the endosperm from each of several seedlings : replace the seedlings and leave them for a day with their roots in water. Fill two paraffined bottles, one with "normal" culture solution : the other with "normal, minus iron." Make several cuts in a plate of cork, passing from the edge to near the centre. Slip the lower parts of the shoots of healthy seedlings each into one of the groove-like cuts, wedge each seedling firmly by means of cotton wool, place the cork over the bottle, the fluid in which is at such a level that it covers the roots but does not reach the shoots. By adopting the precautions mentioned in the appendix, the plants may be kept in a healthy state. Note that the leaves which form in the "normal, minus iron" solution are not green but yellowish or white. In the absence of iron, chlorophyll, the green colouring matter of plants, fails to develope. Paint a part of such a chlorotic leaf with a dilute solution of iron sulphate, and observe that it becomes green in the course of a day or so.

It is a curious fact that though iron is essential for the development of chlorophyll, that substance does not contain iron.

In Nature, certain plants on certain soils become chlorotic and fail to develope properly. It has been shown that this state is due—at all events in many cases—to the inability of the plants to obtain iron from the soil, and that it may be remedied by injecting solutions of iron salts into the stems. It must be borne in mind, however, that not all white leaves are due to the lack of iron. Many plants, *e.g.* maple, ivy, holly, pelargonium, etc., have the habit of forming races with variegated leaves, and, in these cases, we have no reason to think that the absence of colour is due to lack of iron : again, most plants grown in darkness (cf. p. 44) produce colourless leaves. Hence we may suppose that the presence of iron salts, like light, is a necessary condition for the development of chlorophyll.

By referring to the records made of the growth of the sand-cultures, it will be found that, when calcium is absent, the leaves tend to become yellowish, and that, in the absence of magnesium, they often take on a dirty-brown colour. Inasmuch as it has been shown recently that chlorophyll contains magnesium, it should be possible to demonstrate that, in the complete absence of that element, plants do not produce the normal green chloroplasts (chlorophyll grains).

The results of sand- and water-culture experiments prove that the following elements obtained in combination from the soil in the form of salts are essential for the growth and development of flowering plants :—nitrogen, phosphorus, sulphur, calcium, magnesium, potassium and iron ; together with hydrogen and oxygen obtained in the form of water (H_2O). As the result of chemical processes, of which at present the green plant guards the secret, certain of these elements—nitrogen, sulphur, phosphorus, and potash, as well as oxygen and hydrogen—become constituents of plastic food-materials such as the proteins ; and, finally, are incorporated with the protoplasm. Others, such as magnesium, become part of the indispensable chlorophyll grains. Hence we must regard the essential mineral salts as contributing with water to the raw materials of the food.

The precise function of some of these mineral substances is not clear. Some of them, as, for example, iron, seem to condition the formation rather than to enter into the construction of the plant-machinery, others play yet other parts—thus calcium has a rôle in connection with the transport of plastic carbohydrates. For, in its absence, the accumulation of starch in abnormally large quantities in the leaves, indicates that something has gone wrong with the transport service (cf. also p. 185). Calcium occurs also in a combined form in certain layers of the cell-walls.

Our studies of the osmotic properties of cells have taught us that only such mineral substances as are soluble in water can pass into the root-hair cell. That the root has, however, the power of dissolving certain mineral substances, we demonstrate thus :—

110. Pack a large flower-pot half full with coco-fibre : place a slab of polished marble so that one end rests on the coco-fibre and the other on the side of the pot, in such a position that downward-growing roots come in contact with the surface of the marble. Press a layer of coco-fibre firmly into the pot, plant soaked bean seeds, cover them with coco-fibre, and water thoroughly. After some weeks, turn the marble out of the pot and observe that the root-systems of the seedlings have etched upon the slab copies of their outlines. Since marble ($CaCO_3$) is soluble in water containing carbon dioxide (CO_2) in solution, it would suffice for the etching process for the root to produce—as we know it does—carbon dioxide. Whether the root-hairs are able also to excrete other more powerfully solvent substances is a matter of dispute.

That the mineral substances absorbed by the root-hairs are taken up in extremely dilute solution follows from the diluteness of the solutions which suffice for water- and sand-cultures. Indeed, our study of osmosis would lead us to predict that, where the rooting medium contains osmotic substances in any considerable degree of concentration, water absorption is hindered (cf. also p. 128).

We demonstrate the fact that the addition of strong solutions of salts may act injuriously on plants, *e.g.* by plasmolysing the root-hairs, thus :—

111. Raise turnips or oats from seed planted in six pots of ordinary soil. When the seedlings are up, water each of the pots daily with one of the following :—water, 1 %, 3 %, 5 %, 10 % and 20 % potassium nitrate solutions, and determine that the strong solutions exercise a disastrous effect.

Gardeners have long since recognised the fact that a soil may be deficient in some one or other of the essential mineral substances required by plants, and that, even though it may not have lacked such a substance originally, it may come to be poor in an essential mineral, owing to the absorption of this substance by previous crops. They know, also, that animal manures contain relatively small quantities of such substances as phosphates. Hence, in order to restore the fertility of the soil, they apply to it both dung and artificial manures or fertilisers. When a fertiliser is readily soluble in water, *e.g.* nitrate of soda, sulphate of ammonia, etc., care is taken to add it in small quantities, for, otherwise, it is apt to "burn" the plants.

112. This burning effect may be demonstrated by adding powdered sulphate of ammonia to a patch of a lawn infested with daisies. After a few days, the daisies present on the lawn are brown and shrivelled, and the grass itself may be also damaged temporarily. It is owing to the special susceptibility of daisies to damage by sulphate of ammonia that this substance is employed, mixed with sand, as a weed-killer on putting-greens and lawns.

Another matter of interest which our various observations now allow us to explain, at all events in part, is that known as *selective absorption*. Different crops make different demands on the mineral constituents of a soil. Some, like potatoes and root-crops generally, take up more potash salts, others, like leguminous crops, take up more phosphates, and so on. Now, if a substance, say a salt of phosphorus, is taken up by the root and distributed throughout the plant, a moment comes when, if the phosphates in the plant have not in the meantime undergone some change, there is as much phosphate in solution in the cell-sap as in the water of the soil. Hence, automatically, the accumulation of phosphates ceases. If, then, we consider two plants, one which makes use of phosphates largely, the other sparingly, then, though grown in precisely similar soils, the former takes up during its life large quantities of phosphates, the other takes up but little. It will be evident that the behaviour of different crops with respect to amount of absorption from the soil of the essential mineral substances must be a factor in determining the rotation of crops.

It remains to enquire how the dilute solutions of the various mineral substances are absorbed by the root. We have described the plasmatic membrane of the root-hair as a semi-permeable membrane; that is, one which, whilst permeable to water and certain diffusible substances, is impermeable to certain other diffusible substances. We have, therefore, to assume that the root-hair cell, whilst impermeable to the osmotic substances which it contains dissolved in the cell-sap of its vacuole, is, on the other hand, permeable to nitrates, and to the salts of the other elements essential to plants. On this view, we can understand how certain plants take up large quantities of non-essential minerals—just as sea-weeds absorb large quantities of salts of iodine. We suppose that large differences of permeability exist, and that the protoplasmic membrane of the root-hair cells of one plant is permeable to substances which cannot pass the plasmatic membrane of the root-hair cells of other plants.

The study of the behaviour of a plant, the absorbent organs of which find themselves confronted with new osmotic conditions, throws light both on the extent and limitations of the plant's powers of adaptation. Thus, the roots of willows grow in the strongly saline waters of the Dead Sea, and hence the root-hair cells, immersed in the salt water, the osmotic pressure of which is very high, must exert a correspondingly high osmotic pressure in order to retain their turgidity and to effect the absorption of water. Some marine algae withstand a change from salt water to fresh water, a change which would suffice to produce death from " osmotic explosion " in the cells of most plants. Brought into solutions of higher osmotic pressure, the absorbent cells of some plants react by setting up a countervailing osmotic pressure by the secretion of osmotic substances into the cell-sap : other plants reach a similar condition of osmotic equilibrium with the fluid which surrounds them by absorbing osmotic substances from it. On the other hand, plants are not always able to readjust themselves so as to overcome untoward soil-conditions. Thus, a certain number of plants are very intolerant of chalky soils. Among such chalk-shy or calciphobe plants are sphagnum moss, the sun-dew (Drosera rotundifolia), and other plants common in boggy and peaty land ; foxgloves (Digitalis purpurea), rhododendrons, azaleas, etc. The leaves of azaleas and rhododendrons, grown in soil rich in chalk,. turn a sickly yellow colour; whilst the sun-dew and various mosses may be killed outright. It seems probable that though, for some reason or other, an excess of lime is injurious to these plants, their root-hairs are, nevertheless, permeable to soluble salts of lime, and have no power of modifying this property. In the cultivation of azaleas and other chalk-shy plants in greenhouses, gardeners are careful to water them with soft water, and advantage is taken of the susceptibility of various mosses to lime to get ride of them from lawns by watering with lime-water or by ·dressing the lawns with chalk.

Other plants are indifferent to chalk, and yet others (calciphil plants) show a marked preference for chalky soil (Bibliography, 1, 9, 17). Hence, in the cultivation of alpine plants from limestone regions, it is sometimes necessary to use stone containing lime for the construction of rock-gardens.

Our sand- and water-cultures have convinced us of the fundamental importance of the soil to the life of the plant. From it, the plant derives not only its water supplies, but also the mineral substances indispensable to its existence.

In addition to these fundamental relations between plant and soil, there are others hardly less striking, for even in the course of a walk across a small tract of country, it may be seen that the different kinds of plants are not scattered uniformly over it, but group themselves into bands, and that the kinds or types of vegetation characteristic of these several plant-associations are very different from one another (Bibliography, 1, 9).

Over that small tract of country the climate is uniform, throughout it the amounts of rainfall and sunshine are about the same.   Therefore, we are compelled to conclude that the differences in vegetation which its several parts present are due, in some way or other, to differences in the soil.

The same conclusion is reached from a survey of the vegetation of the whole world : woodland and grassland occur the world over, alike in the hot regions near the equator, and in the cool temperate regions. The fact that the species of plants which compose a tropical jungle differ from those which live in an English wood is to be attributed mainly to differences of temperature. Only plants that can withstand a fairly low temperature survive in an English wood, and only plants that can flourish at a uniformly high temperature live in the tropical jungle. But the *tree-type* of vegetation is common to both, and the existence of this type in such different regions is to be attributed to similarity of soil conditions, particularly, as we might guess, to there being a supply of water adequate for the support of luxuriant tree-growth. Thus, the question of the plant's water-supply is one which has many bearings, and is worth careful consideration. That this supply depends ultimately on amount of rainfall—or, in some cases, on supplies of underground water—is evident; though it is also evident, from the facts we have learned (p. 128), that plenty of water in the soil need not necessarily mean plenty of water available for the plant.

Consider, for example, how the vegetation of the land comes in our country to a sudden halt at the sea-shore. None of the plants that flourish a few yards inland are able to establish themselves on the sandy beach. That strip of ground is practically a desert, or at best bears occasional, isolated plants, such as sand-binding grasses (Ammophila arundinacea), yellow horned poppies (Glaucium luteum), sea thistles (Eryngium maritimum), and sea convolvulus (Convolvulus Soldanella). Nor is the dearth of plants along the sea-shore confined to the sandy tracks. In the wet, clayey marshes which mark the estuaries of small streams, the vegetation, consisting of sedges, succulent Suædas and Salicornias (Marsh Samphire, Saltwort), is but poor. Over these marshes, the sea spreads at periods of highest tides, and few plants can live in the soil impregnated with sea salts. Thus, within the space of a few hundred square yards three different types of vegetation may occur; the grassland or woodland, which comes almost to the sea's edge, the semi-desert of the sandy shore, and the salt-marsh. The reason for this grouping of plants according to soil characters is not far to seek. For sandy soils are leaky reservoirs; from them the water drains away almost as fast as it falls as rain upon them. In the salt-marsh also, the casting vote which determines vegetation is given by the soil, for only plants (halophytes) whose roots are tolerant of large quantities of salt can live therein.

Hence a region of uniform rainfall may contain soils which, like those of the meadows or woodlands by the sea, are good water-reservoirs, others which, like the sandy shores, are poor reservoirs—physically dry—and others, like the salt-marshes, which, though they hold much water, withhold it from the majority of plants, and are therefore, though they may contain plenty of water, dry as far as plants are concerned, or, as we say, physiologically dry. Such brief considerations show that a study of the soil is a matter of first importance to the physiologist. To this study—of the physics, chemistry, and biology of the soil—we will now proceed.

Soil is "rotted rock." By the action of rain and running water, frost and other natural agents of destruction, the rocks exposed on the surface of the earth are in part dissolved and carried to the sea by rivers, and in part broken into the fragments of varying size which constitute the soil.

The soil thus formed may accumulate over the rock from which it is derived, or it may be carried by running water for some distance and then be deposited. In the former case, the soil is called a sedentary soil, and in the latter, a transported or drift soil.

Since different rocks are composed of different kinds of minerals, they give rise to soils which differ chemically from one another.

Thus, a limestone disintegrates to form a chalky soil. Rocks rich in felspar (silicate of alumina with potash, soda, or lime) give rise to a clay—a soil consisting mainly of fine particles of hydrated silicate of alumina. If much lime occurs mixed with clay, the soil is termed a marl. From rocks made up chiefly of quartz (silica) the coarser-grained soils, called sandy soils, are formed. Mixtures of sand and clay, in which the decaying remains of plants and animals (humus) have accumulated, are called loams, and constitute the most fertile soils.

The rarer minerals contained in the rocks scarcely affect the general character of the soil, though some are, as we

know, by no means unimportant to the plant. Of these
mineral substances, iron in combination is present in most
soils in sufficient quantities to give rise to characteristic
colours—red, yellow, blue, or grey—which may be so
pronounced as to be recognisable even in soils which,
owing to the humus they contain, tend to be of a blackish
hue.

Fuller information as to the origin of soils may be
obtained from text-books of Geology (Bibliography, 12,
13, 17), and much may be learned concerning the origin
and nature of the soils of a locality by studying them with
the aid of the "drift maps" published by the Geological
Survey.

113. For our first experiment in connection with the
study of the soil, we dig a large hole in a field or garden.
As the earth is turned up and put on one side, note its
depth and also the change of colour as soil merges into
subsoil.

Press a little of the soil between the fingers and note
whether it feels gritty or greasy. If gritty, we know that
it is of a sandy nature; if greasy, that it is clayey.

114. Heat a small quantity of the soil to redness in a
porcelain crucible. When it has cooled, note that, owing
to the humus having been destroyed by burning, the in-
cinerated soil has lost its blackish colour. Repeat the
observations on the subsoil, and thus determine that it is
devoid of humus.

115. Half fill a test tube with a sample of the earth.
Add a little hydrochloric acid : note whether effervescence
occurs. If in doubt, hold the tube to the ear and listen for
a crackling or bubbling sound. Effervescence indicates
that the soil contains carbonate of lime, which, on the
addition of the acid, disengages carbon dioxide.

116* Shake up samples of the soil with water in
tall glass jars or test tubes. Pour off the liquid
from one of the jars almost immediately, and from the
other jars at timed intervals, leaving, however, one jar
undisturbed.

117.* By means of a long pipette, take out samples of
the sediment from each jar and examine it microscopically.
Observe that the larger soil-particles fall first. Mount a

drop of the turbid liquid contained in the last jar, look at it under the microscope and note the minuteness of the particles which it contains. Ascertain how long the water in the undisturbed vessel takes to become clear.

118. If several days elapse before this happens, pour half the contents into a clean vessel and add a little lime-water. Observe that the effect of lime is to precipitate the fine particles, and note that lime is used in agriculture for the purpose of making clay soils more workable.

119. Throw back the soil into the hole (Exp. 113) without treading it down. Note that the earth by no means completely fills the hole. Hence, even though put back with care, it is more tightly packed than before it was dug out, and hence, also, the particles, in their natural position in the soil, are separated from one another by spaces.

120. Plunge a flower-pot, containing a plant not watered recently, in a pail of water : note that, as the water enters, bubbles of air escape from the spaces between the soil-particles.

121. Take two similar plants in pots, *e.g.* geraniums, or chrysanthemums ; submerge the pot containing one plant in water in a pail ; grow the other plant in the usual way. Note that the plant in the water-logged soil fails to flourish. After some time, turn it out from its pot and observe that the roots appear unhealthy and, it may be, rotted. Excess of water means deficiency of air, and hence an unsatisfactory rooting medium (see also Exp. 67).

From the foregoing experiments and considerations, it is evident that a knowledge of the physical, as well as of the chemical, properties of soils is essential to an understanding of the relations which obtain between plant and soil.

We will, therefore, devote ourselves to a brief experimental study of some of the more important physical properties of soils.

122. For this purpose, obtain samples of fairly fine sand or of a sandy soil, and of a heavy clay.

Having dried the soils in an oven, pound and sift them to remove stones, etc. Without pressing down the particles, fill two pots of equal size and of known weight, *e.g.* marmalade jars or pint pots,—one with the sand, the other with the clay. Determine the weights of the equal volumes of sand and clay. Note that sand is heavier than clay, and observe, therefore, that the farmer, when he speaks of a clay soil as "heavy," means that it is heavy to work on account of its tenacious nature.

Next compare the water-capacities of the two samples of soil, that is, the amount of water which can be taken up by an equal volume of each.

123. To do this, fill a graduated burette with water, and run the water from the burette gradually into the vessels containing the sand and clay used in the last experiment. As soon as the soils can absorb no more water, weigh the vessels, and remembering that 1 cubic centimetre of water weighs 1 gram, calculate, from the increased weight, the volume of water absorbed by the sand and clay respectively.

Since both sand and clay consist of solid particles, the water added to the vessels has passed into the spaces between the soil particles, and thus the amounts of water taken up by the sand and clay give us a measure of the total volume of the spaces between the soil particles, or, in other words, of the pore-space of each soil.

124. Turn the wet soil out of the pots, clean, dry, and fill them again with samples of dry, sifted soil—sand and clay as before. Shake down the soil particles by tapping the vessels repeatedly against the table; observe that, as a result of such treatment of the dry soils, the clay particles lie closer to one another than do those of the sand, and that in consequence, the space now occupied by the clay is considerably less than that occupied by the sand.

125.* Mount a small quantity of the sand in a drop of water on a glass slide, and make a similar preparation of the clay. Examine microscopically, and observe that the sand particles are much larger than those of the clay (cf. Exp. 117*). Since a relatively large particle of sand is heavier than a minute particle of clay, sand particles lie more heavily on one another than smaller and lighter clay particles. We thus recognise that the much larger pore-space of the clay soil (Exp. 123) is due, not to the *chemical* properties of clay, but to the fact that it consists of minute particles which lie lightly on one another, and so leave considerable spaces between.

Since they are so small individually, the number of clay particles in a given volume of clay soil is considerably greater than that of sand particles in an equal volume of sandy soil.

Our experiments teach us further that, after a heavy rainfall over fields, some of which have a clay and others a sandy soil, assuming that the layers of soil are of equal thickness, more water is contained in the former than in the latter.

But, as we know, the rain which falls on fields disappears gradually, and we must ask whether the " drying-up " proceeds more quickly in one kind of soil than in another.

If we have at any time observed the sand on the sea-shore, we must have noticed that in summer, soon after the tide has receded, its surface becomes quite dry, whereas the clayey ooze lower down the beach remains for hours glistening with the water which remains on its surface. Evidently, water does not drain away so quickly from a clayey as from a sandy soil.

How marked is the difference between different kinds of soil in this respect we may demonstrate as follows :—

126. Procure two *glazed* flower-pots of equal size, each with a hole in the bottom. Close the holes by means of tight-fitting corks and determine the volume of each pot by measuring the amount of water required to fill it. Remove the corks, dry the pots and place a small filter-paper in each to prevent the soil from escaping through the hole. Fill loosely one pot with dry, sifted sand, the other with dry, sifted clay.

Add water gradually as in Exp. 123 until the soil is saturated and the pots can hold no more. Cover each pot with a glass plate, stand it in a saucer or bowl; weigh, and record the weights. Remove the pots from the saucers and stand them on blocks of wood or retort stands so that the water is free to drop from the holes. Compare the rates at which the water drains away. After some hours weigh the pots again and estimate the relative rates of percolation of the water.

We learn thus that clay holds more water, and holds it more tenaciously, than sand. Whence it follows that, since the water is held in the pore-spaces, a sandy soil becomes dry more quickly than does a clay, and that the latter tends to become water-logged—and, owing to lack of air, sour—more readily than does a sandy soil.

The bearing of these facts on the drainage of agricultural land is evident.

127. When water has ceased to drip from the two pots of Exp. 126, weigh them. Knowing the total amount of water held by known volumes of similar samples of sand and clay (Exp. 123), and knowing the volumes of the glazed pots, calculate the amount of water held by the sandy and clayey soils respectively, after percolation has ceased.

128. Now remove the glass covers from the pots, and, placing the latter where they are sheltered from rain, leave the soil exposed to the air. By weighing at intervals, ascertain that the sandy soil loses water by evaporation more quickly than the clay.

In order to understand how this comes about, we must determine how the water is held in the soil after percolation has ceased. We obtain an insight into the nature of the force by which the water is held on the surface of the soil particles by the following experiment :—

129. Take a series of capillary glass tubes, some of wider, and some of narrower bore : the bore of the largest being about 1 mm., and that of the smallest about .25 mm. Stand the tubes vertically so that their lower ends dip under water, which may be coloured with methylene blue or other dye. Determine the height to which the water rises in the several tubes, and note that, the narrower the bore, the higher the capillary rise. Note also that the upper free surface of the water in the capillaries is not flat but concave, rising higher at the sides where it is in contact with the glass than in the middle ; and observe that the narrower the capillary, the greater is the concavity of the surface of the water contained in it. Since, to lift the water in the tube, work must be done, it follows that the inner surfaces of the tubes exert a pull on the water, and that the narrower the tube the greater the pull. The force holding up the water-columns is called surface-tension, and the water is said to rise in the tubes by capillarity. Tubes so narrow as to exhibit the phenomenon are called capillary tubes. Now we have already discovered (Exp. 123) that the soil is penetrated in all directions by exceedingly fine spaces, and since water, added to the surface, drains away below, it follows that these spaces form a system of inter-communicating channels. With this conception of the soil, it is easy to understand how it is that after water has percolated from a soil the latter still contains a large amount of moisture. We conceive of the water held by the soil as forming films of greater or less thickness around each particle, and of being held thus with considerable force. If a soil consists of extremely fine particles, the number of particles per cubic inch is far larger than that in an equal volume of soil built up of coarse particles. Hence the total surface of the particles of this volume of the fine soil is far greater than that of the particles of the coarse soil.

In illustration, we may quote from Mr. A. D. Hall's book on the soil (Bibliography, 16), the extent of the surface of the particles of a clay soil, a loam and a sandy soil :—

	Pore-space %.	Area of surface of soil particles in sq. ft. per cubic foot of soil.
Fine clay soil -	48	110,500 square feet.
Loam - -	44·1	46,000 ,, ,,
Sandy soil -	32·5	11,000 ,, ,,

From these numbers it is evident that the total surface presented by the particles of a clay soil is about ten times greater than that of the particles of an equal volume of sandy soil, and the number of water-films held tenaciously by surface-tension is, therefore, much greater in the case of the clay soil.

This picture of water in the soil held in the form of films or hollow shells on the surfaces of the soil particles helps us also to imagine what happens as the soil dries by evaporation. As the surface layer exposed to warm air or dry winds loses water in the form of vapour, the water lost is replaced by that attracted from the particles below—for the thicker the film the less firmly it is held : and conversely, the thinner the film the greater the surface-tension exerted upon it by the soil particle. Thus, as evaporation proceeds, there is a redistribution of water in the soil : in other words, there is a *capillary rise*. The finer the system of capillary spaces, the more thorough-going is this upward movement of water from the surface of one particle to that of another. We investigate the rate of capillary rise in various soils, *e.g.* sand, clay, and also chalk and humus thus :—

130. Take a long piece of fairly wide glass tubing and cut it into lengths of about 1 or 2 feet each : fill flower-pots, one with dry, sifted sand, others with clay and powdered chalk respectively (another may also be filled with finely divided peat). Push a tube into each flower-pot so that it is held firmly, pour finely-sifted, air-dry sand into the tube standing in the sandy soil, clay into the tube in the clay soil, etc., tapping gently the sides of the tubes to pack the materials. Plunge the flower-pots in water to their rims and observe, by noting the change of colour of the soil, the rates of ascent of water in the tubes. Contrast the capillary rise in the sand, clay, chalk, etc. Draw or photograph the apparatus at the end of the experiment, and preserve the record in the museum.

The picture of the structure of a soil which arises in the mind as the result of our experiments is that of innumerable particles separated from one another by spaces. Around each particle is a film of water, and the remainder of each space is occupied by air. As the root absorbs water, the films around the soil-particles in the neighbourhood of the root-hairs become thinner. The sur-face-tension which the particles exert draws water from the films around the neighbouring soil-particles. This in turn is absorbed by the roots, which thus obtain water from a large area of the soil. In course of time, if no fresh supplies of water reach the soil, the osmotic pressure set up by the root-hairs is counterbalanced by the surface-tension exerted by the soil-particles, and the root ceases to absorb water.

The soil, however, loses water not only to the plant, but also by percolation and by evaporation. Hence the horticulturist, who cannot always supply water to the soil, has often to take steps to preserve the water in a drying soil.

This may be effected, for example, by spreading lightly over the surface of the soil a layer of fine particles, *e.g.* of finely-sifted, dry soil. By this means, the continuity of the network of capillary tube-like spaces, which extend from the depth of the soil to the surface, is interrupted at the surface, and, evaporation of water from the top layer being checked, the capillary rise of water from the deeper layer ceases. It is as though an infinite number of miniature bungs were put in the necks of as many capillary bottles. On the other hand, if the fine surface-layer is pressed down firmly, for example, by rolling, and its particles thus brought into close contact with those of the layer beneath, capillary continuity is re-established, the particles of the top layer are able now by surface-tension to obtain fresh supplies of water from that surrounding the particles with which they are in contact, and hence, as the water of their films evaporates, there is established an upward movement of water from particle to particle. Again, if, instead of finely-divided soil, a loose layer or mulch of material, such as straw, or grass cuttings, or leaves, is placed on the surface, evaporation is checked. This layer, not in capillary continuity with the soil-particles, may itself become dry, but cannot withdraw the water from the soil below. Thus, the capillary rise being stopped, the water in the soil is conserved.

It is of interest to observe how agricultural and horticultural practice have for one of their main objects the control of the water of the soil. Nor, when we realise the water-requirements of plants, is it surprising that this should be the case?

The rainfall of this country, though spread over many days, is not great. In the neighbourhood of London, the annual rainfall is about 24 inches. Of the rain which falls on a field, some runs off the surface, much passes by percolation to the deeper layers of the soil and is carried away by drainage; much is lost by evaporation during those parts of the year when the fields are unoccupied by crops. Only that which remains is available for the plants.

That there is apt to arise a shortage of water, and that the crop is likely to suffer, is evident when we consider how considerable are the water-requirements of various crops. Thus, the amount of water taken from the soil by a crop of wheat is equivalent to a rainfall over the whole wheat-field of no less than six inches. Other crops are yet more thirsty; mangolds, for instance, absorb from the soil the equivalent of 10·6 inches, or something like half the total, average annual rainfall in southern England.

It is in consequence of the annual risk of water-famine among crops that tillers of the land are at such pains to interfere with nature in the interest of the crops they cultivate. Thus, autumn ploughing, by breaking up the surface layers of the soil, increases its water-holding capacity and, as the frosts of winter cause the clods to crumble, the water is conserved by the interruption of capillary continuity.

The practice of hoeing, resorted to by all good gardeners, has a doubly beneficial effect : in the first place, it keeps down weeds, which compete with the cultivated plants for the water or other materials in the soil; in the second place, by breaking the upper layer of earth, a fine tilth is established and water is conserved in the soil.

How important are the effects of hoeing may be judged from the following figures, which give the loss of water from hoed and unhoed areas of soil, deduced from an actual experiment :—

Daily loss of water from cultivated soil, 14·5 tons of water per acre.
    ,,   ,,   ,,   uncultivated ,,   17·6   ,,   ,,

In this experiment, soil cultivated to a depth of 3 inches was sampled at intervals to a depth of 6 feet. Control samples were taken simultaneously from adjoining land with a firm surface.

There was thus during the 49 days of the experiment a saving of water corresponding to 1.7 inches of rain.

We must not bring these studies of the soil to a close without considering the soil, albeit in the briefest manner, from another aspect, namely, the biological.

The soil is rich in many forms of life. Beside the flowering plants, it bears ferns and toadstools. Sometimes a green scum of minute algae may be seen on its surface, and by appropriate methods a rich flora of non-green organisms, fungi, bacteria, and also a fauna of minute, unicellular animals may be discovered to have their homes in the upper layers of the earth.

Though, however, the parts played by the soil-flora and fauna in determining fertility are of the greatest importance, we can consider them here in the light of but one example.

131. Repeat Exp. 106, but mix with the sterile sand in one of the pots about an ounce of ordinary garden soil.

Water this pot, and those containing sterile sand only, with culture solution lacking nitrogen (Appendix A). Observe that the peas growing in sand to which the earth was added, though they may remain puny for a while, subsequently grow as vigorously as those planted in garden soil. Those in sterile sand remain small. After some weeks, turn the plants out of their pots and observe that whereas those which have grown vigorously have curious nodules or swellings on their roots, those in the sterile sand have none. Keep the specimens for the museum.

132. Germinate seeds of gorse (Ulex europæus), some in sand sterilized by heating in a hot oven, and others in ordinary soil. Observe the development of the seedlings and the relation between their growth and the presence of nodules. Dig up any leguminous plant from the field or garden (vetch, lucerne, lupine, etc.), and observe its root-nodules. From laborious investigation of these root-nodules, it has been proved that they are the result of an infection of the root by a definite micro-organism, Pseudomonas radicicola, and it has been demonstrated that this micro-organism, even when isolated in artificial cultures in the laboratory, has the power of obtaining its nitrogen from the free nitrogen of the air. This it does also when it lives in partnership (symbiosis) with the roots of leguminous plants. These plants, however, not only tolerate the micro-organism which multiplies in the cells of their roots, but eventually digest it, and so secure the nitrogen which the micro-organism has amassed. We thus learn that micro-organisms exist in the soil which have the power to "fix free nitrogen," that is to bring the nitrogen gas of the atmosphere into such combinations as serve plants for food-materials.

We may understand how all-important this fact is when we remember that the amount of combined nitrogen (nitrates, etc.) in the soil at any given time is but small ; that plants are constantly taking toll of it ; that, in the decay of plants and animals, the complex proteins are broken down by various micro-organisms, in ordered stages, to form ammonia and ultimately nitrogen. The nitrogen so formed escapes into the air, as may also some of the ammonia. There is thus, as it were, a constant leakage of nitrogen from the living world. By the agency of the nitrogen-fixing bacteria, some species of which occur in the nodules of leguminous plants and other kinds of which live exclusively in the soil, this escaped nitrogen is once more brought into combination and rendered available for the nutrition of the higher plants.

It is more than probable that the general shortage of nitrogen in earth and sea has led to other relations between organisms as remarkable as that between leguminous plants and Pseudomonas radicicola.

133.* Thus, when we examine with lens and microscope the young roots of forest trees, *e.g.* pine, oak, hornbeam, etc., and also those of heath plants, we discover that, instead of root-hairs, the young roots are covered with mantles of fungous threads (mycorhiza).

It is not unlikely that, as a result of this symbiosis, the flowering plants which take a part in it obtain increased supplies of nitrogen. The lichens which grow on rocks and tree stems and other situations represent symbiotic partnerships between fungi and algae. Not a few animals, of which the green hydra is an example, are associated always with green algae. It is possible that all these strange symbiotic unions have for their significance the solving of the problem of obtaining adequate supplies of nitrogen.

CHAPTER IX.

The absorption and loss of water by the plant. The water-requirements of various types of plants :—hygrophytes and xerophytes. The process of the transpiration of water by the leaves : the structure of the leaf in relation to this process : the part played by stomata : the opening and closing of stomata and the conditions under which these movements occur. Apparatus for measuring rate of transpiration— (Potometer).

Now that we have completed our studies of the properties of the soil and the bearing of these properties on plant-life, we have to discover what becomes of the water and mineral substances which are taken up by the root-system.

As a preliminary to this work we proceed to ascertain experimentally the amount of water absorbed by the roots of various plants. The most accurate method is to measure absorption directly ; a simpler way is that given in Exp. 137.

134. To employ the first method, germinate maize or other seeds used in water-cultures (Exp. 109) so that the roots of the seedlings grow into a normal culture solution. Prepare two tall glass vessels, each provided with a graduated side tube (Fig. 23). Fill the vessels with normal culture solution to within an inch or so of the top, transfer a well-developed seedling to one of the vessels so that its roots are in the solution. Fix its stem in a split cork which fits into the neck of the vessel. By means of paraffin or wax, (Appendix A) make air-tight the junctions between cork and glass, and cork and seedling. Plug the side tube with a wad of cotton wool to keep out dust. Wrap black cloth round the vessel in order to exclude light from the roots and thus to prevent the growth of green algae in the culture solution; but leave the side tube uncovered. Fix a thermometer by rubber bands to the side tube. Proceed in a similar way with the other vessel. Label the vessels A and B, place them in a good light, and continue, for as long a period as possible, to take daily records, by reading the level of the fluid in the side tube,

FIG. 23.—APPARATUS FOR MEASURING ABSORPTION OF WATER BY PLANTS.

of the amount of water absorbed by the plants. Note each day the temperature of the air and also the state of the weather, fine and bright or wet and dull. Add fresh culture solution whenever necessary.

Determine the total amount of water absorbed by A and B; plot the daily amounts on squared paper, and draw "curves" of the amounts of absorption. Observe any relation which appears to exist between weather and amount of absorption. Since the volume of water absorbed by the plant during the course of the experiment is many times that of the plant itself, it follows that the water is either decomposed or else given off by the stem or leaves. The readiest method of demonstrating that water is given off from the plant consists in covering it with a bell-jar and observing that drops of water condense on the inner surface of the glass. Or we may use our plants A and B for the purpose thus :—

135. Weigh the apparatus containing plant A. After two or more days weigh it again. Determine that the loss of weight of the apparatus corresponds approximately to the weight of water absorbed by the root during the period of the experiment.

Next demonstrate, in the following way, that the loss of weight recorded in the preceding experiment is due to loss of water by the plant.

136. After weighing apparatus B and recording the level of the liquid in the side tube, close its side tube with a cork and cover the plant in B with a large bell-jar, under which is placed a weighed vessel containing calcium chloride. The rim of the bell-jar must be sealed by wax to a flat block of wood or to the table, in order to prevent the calcium chloride, which is very hygroscopic, from absorbing water from air entering the bell-jar from outside. After 12-24 hours remove the bell-jar and also the cork from the side tube; weigh B and weigh the vessel containing the calcium chloride; by reading the level of the liquid in the side tube, determine the amount of water absorbed by the plant during the course of the experiment. Observe that the loss of weight of B is about equal to the gain in weight of the calcium chloride. Since the increase in weight of the latter is due to absorption of water, we infer that the loss of weight by the plant is due to loss of water. Hence it follows that the water absorbed by the root-system passes up the stem and is given off into the air in the form of water vapour, leaving behind in the tissues of the plant the mineral substances which it held in solution. Evidently therefore one important function of the water absorbed by the root is to act as a vehicle for the carriage of the mineral salts from the soil *via* the root to the shoot. The fact that the loss in weight of plant B (Exp. 135) is approximately equal to the weight of water absorbed by the root during the same time does not mean that no other processes involving change of weight are going on in the plant, but that the amount of change in weight due to such processes is, during the short time of the experiment, small in comparison with that due to loss of water.

The second and simpler method which we use to estimate absorption is based on the fact, which we have just demonstrated, that the amount of water absorbed in a given time by a herbaceous plant is about equal to the amount given off as vapour.   It consists in putting pot-plants under such conditions that evaporation from soil and pot is prevented, and then weighing the plants at intervals.   Loss of weight is due mainly to loss of water : amount of loss of water is approximately equal to amount of absorption : therefore loss of weight gives a rough measure of amount of absorption.

137. The apparatus is made as follows :—Choose two leafy sunflower or dwarf bean plants grown in small pots. Procure marmalade or other jars, each large enough to hold one of the pots.   After watering the plants thoroughly, let the superfluous water drain away and stand the pots on pieces of tile placed at the bottom of the jars.   From a sheet of American cloth cut out two circular pieces to serve as covers for the jars.   Make a cut in each piece extending from the circumference to the centre.   At the centre, cut out a small circular piece to allow of the stem passing through.   Place each cover in position.   Draw the cut edges together by means of stitches, turn the cloth down over the rim of the jar and tie it firmly by means of several turns of stout thread.   Or the same end—that of preventing evaporation from the soil, etc.—may be served by getting a tinman to cut out for each jar two semi-circular pieces of tin, each notched to enclose the stem. Two such tin plates, when fixed by wax to one another and to the edge of the jar, make a good cover.   Label the pots A and B.   It will be necessary from time to time to remove the covers and add measured quantities of water. From the series of daily records of the loss of weight of

the plants, calculate the total amount of water which they
have given off during the time of the experiment (cf. Exp.
136).

We determine next that the water given off escapes
mainly from the leaves.

138. After weighing the two vessels with their plants
(using the apparatus either of Exp. 134 or that of Exp. 137),
remove about half of the leaves from plant B. Weigh
the apparatus again, replace it and take a daily or 48-
hourly record of its loss in weight. Whilst the cut-off
leaves are fresh, determine their area (Appendix B). Note
that the rate of loss of water by the plant from which
leaves have been removed is considerably less than it
was before. Remove the remaining leaves and estimate
their area. From the records obtained, calculate the rate
of loss of water *per unit of area,* say per 10 sq. cm. of leaf
surface. We thus demonstrate that the amount of loss of
water, or, as we say, the amount of water transpired, is
roughly proportional to extent of leaf surface, and infer
therefore that the water escapes mainly from the leaves.
Since a small quantity of water is also given off from the
stem, even after the precaution of covering, *e.g.* with
collodion, all the wounds formed by removal of the
leaves, it may be necessary, in making this calculation,
to deduct the amount of water given off by the leafless
plant from the amount given off by the plant with half
its leaves, and from that with all its leaves intact. In
the foregoing experiment, plant A—with its leaves
intact—serves as a control. We know the relative rates
of transpiration of A and B over a period of many
days. We know the rate of transpiration of A during the
period of the experiment. Hence we can calculate approximately the amount of water which B would have lost if its
leaves had not been removed, and we are able to compare
the rate of loss of B with all its leaves with that of B with
half, and with none of its leaves. By continuing to record
the transpiration of plant B for some days, we may
observe that though it is very much reduced, some loss of
water occurs after all the leaves have been removed, and
conclude that, though the leaves are special organs for
allowing of the escape of water vapour, other parts, par-

ticularly the delicate green parts of the stems, may also give off water vapour.

Plant B is now useless for further experiment, but plant A should be grown for as long a period as possible, and regular records of the rate of transpiration made, tabulated, and the record-chart kept in the museum.

Having gained precise information with respect to the large amount of water taken up and given off by plants with thin delicate leaves like maize, sunflower or bean, we proceed to find out whether plants with other types of leaves are equally spendthrift of water. For this purpose, use any pot-plants the leaves (or stems) of which are thick and fleshy, *e.g.* Agave, india-rubber plant (Ficus elastica), any cactus with fleshy stems, a Mesembryanthemum, or, failing any of these, gorse (Ulex europæus) grown in pots.

139. Having fitted up a broad, thin-leaved plant (A), *e.g.* sunflower, in a marmalade pot A (Exp. 137), fit in apparatus B one of the just-mentioned plants. Determine, by weighing daily, the rates at which A and B lose water. After some days, determine the areas of the leaves and stems of the two plants and calculate their rates of transpiration per unit of area.

The following is a record of the rate of transpiration of two plants placed under similar conditions for twenty-four hours. It shows that the rate of transpiration per 100 sq. cm. of surface was 32·7 times as great in the case of Hydrangea hortensis, a plant with relatively large and thin leaves, as it was in that of Opuntia cylindrica, a cactus with a fleshy green stem and no leaves.

	Transpiration per 100 sq. cm of surface.
Hydrangea hortensis -	6·54 gm.
Opuntia cylindrica - -	0·20 ,,

We learn from these experiments that different kinds of plants have very different water requirements. Some lose water and hence if they are not to wither, also take up water, much faster than others. Not only is the loss of water per unit of surface much less in succulent and leathery-leaved plants than in plants with delicate leaves, but also the total leaf- (or green stem-) surface of the former is much less than that of the latter. Since such succulent plants live generally in dry situations, we may infer that the reduction of surface, the succulence of the leaf or stem, the thick cuticular covering of leathery leaves, the matted hairs common on such Alpine plants as the Edelweiss (Leontopodium alpinum) are so many adaptations making for water economy. Such plants find water either hard to get (desert-plants, alpines) or hard to keep (plants of hot countries, agaves, cactuses, etc.) or both. Plants which manage to live in such situations do so by definite adaptation of their root-systems and shoot-systems : the root-system of desert-plants is often of extraordinary length— the roots, as it were, going in search of water : the shoot-systems display the most diverse adaptations serving to reduce the amount of water given off from their surfaces. These water-economising plants are known as *xerophytes* in contradistinction to *hygrophytes*. The latter are as extravagant in the amount of water which they absorb and give off as the former are niggardly.

Among British plants, note that those of sandy soils and other dry situations are xerophytes, *e.g.* the Scotch fir (Pinus sylvestris), gorse (Ulex europæus), broom (Cytisus scoparius), heaths and various grasses which grow in their company.

140. Specimens of these xerophytes and also of the commoner hygrophytes—broad-leaved trees, etc.—should be collected, compared, and kept in the museum.

Having learned something of the water requirements of various kinds of plants, we next enquire how the water and salts absorbed by the roots pass to the leaves, and how the leaves get rid of the large quantities of water which reach them. We will first consider the latter question.

Clothes hung upon a clothes line in the open air dry soon if the weather is bright and warm. Since, in the course of Exps. 134 and 137, it has surely been noted that plants lose water more quickly on a bright or warm day than on a dull or cold day, it might be supposed that the loss of water by the leaf is, like that of clothes hung out to dry, due solely to evaporation. But, if this were the case, the plant would be altogether at the mercy of its surroundings. In hot, dry weather, when evaporation is high and absorption, owing to dryness of soil, is low, the plant would wither, wilt and die.

So it may, if the drought is prolonged; but not with the inevitableness of the drying of clothes in bright weather. From their behaviour in spells of dry weather, it looks as though plants did not give way to drought without a struggle, and that, though dry air causes evaporation from clothes and plants alike, the latter, unlike the former, have, to some extent at all events, means of preventing extreme losses of water from their tissues. We can imitate this resistance of the plant to the last stages of desiccation by the following experiment :—

141. Weigh two large saucers or basins, fill one with a known volume of water, the other with an equal volume of fairly strong solution of sugar. Weigh the vessels containing the water and the sugar-solution. Expose them in a warm place, and determine, by weighing, the rates of loss due to evaporation. Note that, as the sugar-solution concentrates, its rate of loss of weight falls off very considerably. Consider now a vegetable cell : within its protoplasmic (plasmatic) membrane is a vacuole containing sugar, salts, etc., in solution. Enclosing it is a wall of cellulose. Both cell-wall substance and protoplasm contain imbibed water. Imagine a tissue made up of a number of such cells exposed to a dry atmosphere : the cell-wall loses water by evaporation and withdraws water from the protoplasm, which in turn withdraws water from the cell-sap. The cell-sap becomes more concentrated, and hence its osmotic pressure, *i.e.* its attraction for water, is increased. As evaporation proceeds, not only is the imbibed water remaining in cell-wall and protoplasm held more strongly, but also the concentrated cell-sap, unless it is able to withdraw water from neighbouring cells and to become dilute again, offers, like the sugar-solution of our experiment, more and more opposition to evaporation. Thus we understand one way in which the plant may resist extreme desiccation. Before deciding, however, that this is the only method, we must look at the leaf itself, for its structure may throw further light on the subject.

On inspecting various leaves, we observe that in many of them there are marked differences between their two surfaces.

142. Compare, for example, from this point of view,

the following :—beech, elm, holly, tulip, hyacinth, laurel,
etc. Note that in leaves which, in their natural position,
lie more or less horizontally, the under side is paler than
the upper side. The surface exposed to the direct rays of
the sun appears to be less delicate than the other, and
looks as though it would allow water to evaporate from it
less rapidly. We prove that this is the case thus :—

143. Take two similar leaves of the india-rubber plant
(Ficus elastica) or the laurel (Prunus Laurocerasus). Slip a
piece of thin rubber tubing over the stalk of each leaf.
Turn back the tubing, and, by means of wire, bind the loop
thus made so as to block the hole in the tube and to make a
convenient hook. Smear with vaseline the lower side of
one leaf and the upper side of the other. Weigh each leaf.
Hang up the leaves in a room, and determine, by weighing
at daily intervals, the rates at which the leaves are losing
water.

144. Not all leaves, however, show this marked difference between under and upper side, and if the above experiment is repeated on the leaves of beech or broad bean, using
in these cases *similar branches* with equal numbers of leaves,
no marked difference between under and upper sides with
respect to resistance to evaporation will be found. This,
however, does not affect the fact shown by Exp. 143, that
plants which require to reduce evaporation, and yet, at the
same time, are obliged, for other purposes, to expose a considerable leaf-surface to the light, produce leaves the upper
surfaces of which hinder evaporation more than the lower.
Microscopic examination will help us to learn more as to
the nature of the differences between the upper and lower
surfaces of leaves.

145.* Mount a small piece of a leaf of the frogbit (Hydrocharis Morsus-ranae) or water starwort (Callitriche verna)
in a drop of water, and examine it with low and high
powers of the microscope. Observe that, in one or other
or both surfaces, there are large numbers of extremely
minute holes. These holes or *stomata* cannot be seen in
thicker leaves when whole pieces are mounted. Therefore,
in order to examine ordinary, thick leaves for stomata,
strip off, by means of fine forceps, or cut with a razor,
pieces from the upper and lower surfaces of the leaves of the

plants mentioned below; mount each piece in water on a slide, and determine that, in the india-rubber plant and laurel, stomata occur only on the under side of the leaf, whereas in the bean, beech and tulip they occur on both sides. The fact that the surfaces of leaves are not continuous but are interrupted by innumerable openings throws a new light on the results of Exps. 143 and 144, and at once suggests the idea that water-vapour may escape not only from the general surface, but also through the stomata. This becomes the more likely when we discover, by the following experiment, that the stomata are in communication with spaces running between the groups of cells of which the soft tissues of the leaf are composed.

146. Attach firmly by wire a small piece of rubber tubing to the stalk of a delicate leaf, *e.g.* marsh marigold (Caltha palustris) or wild arum (Arum maculatum). Immerse the blade in water and suck at the open end of the tube. Note that the leaf becomes darker green owing to the entrance of water through the stomata. If the surface of the leaf is first smeared with vaseline, the stomata are blocked and no water enters. Or, if an air pump is available, connect it with the side tube of an Erlenmeyer flask (Appendix B) containing water. Place a delicate leaf beneath the surface of the water. Having closed the opening of the flask with a rubber cork, exhaust the air by means of the pump. Note that, as the pressure in the flask is reduced, bubbles of air escape from the cut end of the petiole and water enters through the blade. Hence the stomata communicate with spaces—intercellular spaces—which run between the cells of the leaf. Hence also evaporation is possible not only from the surface tissues of the leaf, but also from the cells of deeper tissues abutting on the air spaces. For, if the air at the surface of the leaf is dry, water-vapour passes by diffusion from the air spaces through the stomata. Thus the air of the spaces becomes drier, and, in consequence, water evaporates from the cell-walls; in short, the internal cells neighbouring on the air-spaces lose water by the same process as that described for the cells on the surface of the leaf (p. 150). We are not, however, to suppose that the purpose of this system of spaces and of stomata is necessarily to facilitate transpiration. Their main function, for aught we know at present, may be of quite another kind, *e.g.* for allowing air to circulate freely in the leaf. But whatever their main function may be (see p. 161), the facts remain, that they are there, and that, therefore, they cannot but serve as channels through which water-vapour, given off by the cells abutting on them, escapes into the air. That water-vapour does escape from the stomata we demonstrate thus :—

147. Soak strips of filter- or blotting-paper in a 10 % solution of cobalt chloride. Dry them in a desiccator, and when not in use, keep them there or in a closely stoppered bottle. Expose a strip to moist air, *e.g.* breathe on it, and note that it changes colour. Cut off a leaf; place on either side of it a strip of cobalt chloride paper covered by a sheet of thin glass or mica; bind the glass close to the leaf by means of rubber-rings, lay the leaf on a clean table and cover it with a bell-jar or glass dish. Observe the change of colour on one or both sides of the leaf and determine that stomata occur, in the one case, on one side, and, in the other, on both sides. Instead of being cut off, the leaf may be left attached to the plant during the course of the experiment.

148.* If a micrometer is available, determine (1) the number, (2) the size of the stomata either on strips from the surfaces of the leaves or on the whole leaves of Exp. 145. It may thus be shown that the number of stomata in a leaf is almost incredibly large; ranging generally from 100 to 700 per sq. mm. of leaf surface: *i.e.* from 50,000 to over 350,000 per square inch. The size of each stoma is as small as the number of stomata is large: on the average a stoma is about ·0006 mm. in diameter; *i.e.* ·000024 inch.

Our discovery that the surface of a leaf is riddled with holes of microscopic size is a fact which has evidently to be taken into consideration in studying the mode by which water is given off by green plants. As we have proved, the water lost by the leaf escapes in large measure not from the general surface but through the stomata. Now, when we were examining these pores (Exp. 145), we could scarcely fail to notice that each of them is bounded on either side by a cell, the shape of which is different from that of the other cells of the surface layer (epidermis) of the leaf. The cells which encompass the stomatal space are called guard-cells. In surface view, each guard-cell appears more or less sausage-shaped. The guard-cells have the further peculiarity that, whilst they are connected with one another at their ends, they are separated along the middle line, the cleft between them constituting the stoma.

Having obtained an idea of the nature of stomata and their guard-cells, we return to our problem. Is loss of water by the leaf merely a matter of evaporation, or has the plant—as we suspect—some power of regulating its loss? The only parts of the apparatus concerned in the giving off of water likely to be under the control of the plant are the stomata. Now, stomata are holes—slits between cells. If the plant could control the size of the stomata, then the amount of water-vapour escaping through them might perhaps be reduced in amount. But if the stomata are to change in size, they can do so only by change in the shape or size of the guard-cells. Therefore, we have to ask, are the guard-cells fixed in position or are they moveable?

We know already that a cell, when it becomes turgid, increases in size, and that, when it is plasmolysed, it shrinks. We proceed to ascertain whether any change involving change in the stomata, occurs when guard-cells are rendered turgid or when they are plasmolysed.

149.* Repeat Exp. 145 on surface-sections of leaves of Tradescantia or other convenient plants.

The leaves of the following plants serve well for observations on stomata, as the epidermis may be removed easily by the aid of forceps. Primula sinensis, Primula obconica, Tropaeolum majus, Pelargonium zonale, Helianthus annuus, Tulipa, Vicia Faba.

Note that whereas, when they are in water, the stomata are large or, as we may say, open, when a 10 % salt-solution is run in they become small, that is, they close : note further that this change is effected by a change in shape and size of the guard-cells. A complete understanding of the mechanics of this remarkable movement of guard-cells involves a detailed study of their structure ; a study upon which we cannot now embark (Bibliography, 5, 6, 8).

Knowing that stomata are capable of being increased or decreased in size, we next enquire whether they actually change in size during the life of the plant, and if so, what are the conditions which determine the movements of the guard-cells by which the opening and shutting are effected? We therefore investigate the state of the stomata in plants which have been subjected to diverse conditions, *e.g.* bright light, darkness, dry air, moist air, lack of water at the root, etc.

To make the observations, we require a better method than that employed in Exp. 145, though the plants used in that experiment will serve admirably. The method we use is as follows :—

150.* Expose growing pot-plants for an hour or so (1) to bright light, (2) to darkness, (3) to dry air (see Exp. 136), and also cut off leaves and allow them to wither. Throw the leaves treated in these ways into bottles containing absolute alcohol; stopper and label each bottle; expose the bottles to sunlight till the leaves are colourless. Transfer the leaves to another bottle of absolute alcohol. After an hour, take them out and plunge them into a bottle containing xylol. When the leaves are transparent, mount small pieces in cedar wood oil and examine under low and high powers of the microscope. Conclude from the experiment that the stomata of the leaves of many plants close in darkness, in dry air, when the soil is dry, and when the leaves wither, and that, on the contrary, they open in bright light and in moist air.

It is evident that certain of the conditions which make for the closing of stomata are conditions under which loss of water by evaporation tends to become excessive, *e.g.* dry air. On the other hand, it is *not* evident that change from light to darkness—apart from attendant temperature change—results in a reduction of the rate of evaporation. Let us ascertain, therefore, whether change from light to darkness produces any change in rate of loss of water by the plant. If it does, we shall be justified in concluding that the opening and closing of stomata serve, in some measure, to regulate transpiration. We shall not, however, be justified in concluding that this is the primary significance of these movements. For, as we have suggested already, the stomata are not merely concerned with transpiration, but also with exchange of gases between plant and air, and it may well be that the opening of the stomata in the light and their closing in the dark are concerned primarily with the regulation of gaseous exchange, and only indirectly with transpiration.

In order to investigate the effect of external conditions, light, darkness, etc., on the rate of transpiration, we require a more delicate apparatus than that used in Exp. 134 or Exp. 137; though for certain experiments (Nos. 153 and 154) either of those may be employed. The principle on which our more delicate method is based is as follows :—

151. The rate of transpiration by an herbaceous plant is about equal to the rate of absorption (p. 145). Since the water absorbed passes from the root to the leaves, it must pass through the stem. Therefore, if we connect the stem of a living plant with a narrow tube, we may be able to see the water passing up the tube and to measure the rate at which it passes. Such an apparatus, called a potometer, may be of one of several forms. That shown in Fig. 24, and designed by Professor Farmer, consists of the following parts :—A wide-mouthed bottle which is fitted with a rubber cork pierced with three holes; the middle hole being just large enough to admit of the passage of the stem of a fair-sized leafy branch. A thistle funnel provided with a stop-cock is fitted into one of the other holes, and into the third, a bent tube of narrow bore is passed, so that one end is flush with the inner surface of the cork.

Another type (Fig. 25)—the original form devised by Francis Darwin—consists of a glass tube with a side limb, to the free end of which a short piece of stout rubber tubing is attached securely by wire. The branch of the plant to be used in the experiment is passed through the rubber tube into the limb of the glass vessel and fixed firmly by wire wound round the rubber tubing. When in use, the upper end of the straight limb is closed by a well-fitting cork, and the lower end is fitted with a rubber cork with one hole, through which passes a glass capillary tube of about five or six inches in length. In setting up either apparatus, the following precautions are to be observed :—

(1) Cut off the branch under water, *e.g.* by bending it down beneath the water contained in a basin or pail.

(2) Keep it for some hours with its cut end under water, *e.g.* if a laurel branch is used, cut it off in the evening and use it the following morning.

(3) Before it is put into the potometer, remove a short length ($\frac{1}{4}$ inch) of the cut end by means of a sharp razor (making the cut under water).

(4) When inserting the branch, fill the potometer with water and submerge it in a sink or pail filled with water.

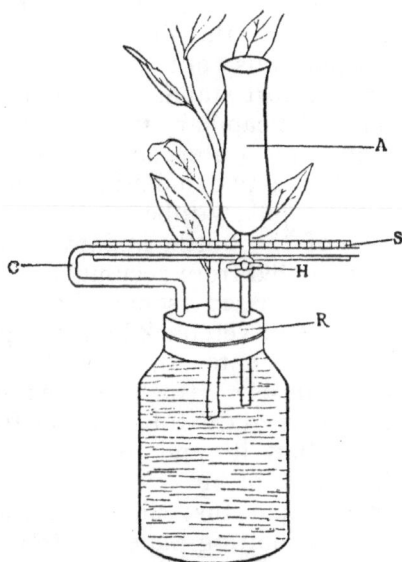

FIG. 24.—POTOMETER. (AFTER FARMER.)

A, funnel with stop-cock (H); C, glass tube of narrow bore; S, scale; R, rubber cork.

(5) Fill a tumbler with water, insert the cut end of the branch in the tumbler, carry it to the sink, submerge the tumbler and transfer the branch to the potometer. If the bottle-potometer is used, submerge the rubber cork and pass the cut end of the branch through the middle hole so that, when the cork is inserted in the neck, the branch projects into the bottle. Press the cork into the bottle, close the stop-cock, pour water into the thistle funnel, and

lift the apparatus out of the water. See that the junction between cork and branch is good : if not, dry the surface and make the joints air-tight with wax (Appendix A). Place the apparatus in a good light and fix a scale behind the narrow tube, or make on it india-ink marks at a distance of three or more inches from one another.

If the tube-potometer is used, submerge it, and, keeping the cut end of the branch from contact with air, pass the branch through the rubber tubing; wire it by many tight turns of thin wire passed round the rubber tubing. When the joint has been made good, push in the cork bearing the capillary tube, cork the upper end of the straight limb, lift the apparatus out of the water, and fix it in the clamp of a retort stand. Adjust the apparatus so that the free end of the capillary tube just dips into a small glass vessel containing water and standing on a small block of wood. Make two india-ink marks about three inches apart on the capillary tube.

One advantage of the bottle apparatus is that, by its use, air may be prevented from accumulating in the vessel; for when air has passed some distance along the narrow tube it may be driven back by turning the stop-cock and allowing water to enter the vessel from the thistle funnel. When not in use, attach by means of a piece of rubber tubing, a bent glass tube to the free end of the narrow tube, and allow its end to dip in water in order to prevent the entrance of air.

To use the tube apparatus, slip away the wood block, remove, by means of blotting-paper, a drop of water from the free end of the capillary tube so as to admit a small bubble of air, replace the vessel and block so that water again enters. Time, by means of a stop-watch, the passage of the air bubble from the lower to the upper mark. Note the time. Admit another air bubble, repeat the timing, and so on. If the branch is losing water rapidly, the rate of passage of the bubbles is very fast, and a few minutes suffice to give 10 or 12 readings.

When the bottle-potometer is used, the end of the water column in the narrow tube is followed and the time it takes to pass from mark to mark recorded. It is driven back by opening the stop-cock and admitting water to the bottle. Thus a series of readings may be obtained. An average of the series gives the rate at which water is passing

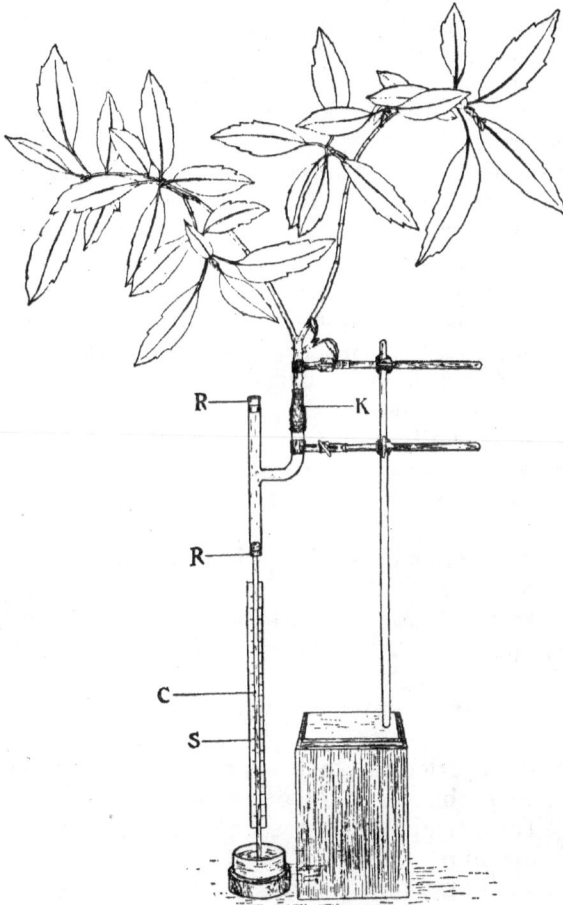

FIG. 25.—POTOMETER. (See p. 156.)
C, capillary tube ; S, scale ; R R, rubber corks ; K, rubber tubing.

into the cut end of the stem ; and since, as we have shown, this rate is, generally, proportional to the rate of transpiration, our potometer gives us a measure—albeit an indirect one—of the rate of loss of water by the leaves. The chief value of the apparatus, however, lies in its service for quick comparative records : in such comparisons, we may assume that, if under one set of circumstances, the bubbles pass, on the average, twice as quickly as under another set, the rate of transpiration is twice as great in the former as in the latter circumstances. The records of rate of transpiration now to be taken should be accompanied by temperature-records.

152. Expose the apparatus in the window of a room; record the air temperature, take a series of readings, and determine the average rate at which water is passing along the graduated tube. Transfer the apparatus to the open, and set it in a breezy, shady place; after a short time, take a series of readings. Expose it in the open to bright sunlight, and, after an interval, make another series of readings. Replace it in its original position, and, after a quarter of an hour, take readings. Put it in a dark place, leave it for about half an hour, and make a series of records. The plant may be screened from the light (*e.g.* a candle) used in reading the rate of passage of the bubbles.

153. Fit a small leafy branch in the tube-potometer; having taken a series of readings in ordinary air, cover the branch with a bell-jar, provided, if not otherwise tall enough, with a "flounce" of glazed calico. Make the air thoroughly moist, *e.g.* by syringing water under the bell-jar (avoiding wetting the leaves). Record : remove the bell-jar, and after an interval take a series of readings in ordinary air.

154. Having dried the bell-jar, replace it over the plant, and put under the jar a large dish containing calcium chloride. Take readings, one series at once, another after about half an hour.

The conclusions which we draw from the results of the foregoing experiments are as follows :—In the first place, conditions which favour evaporation favour transpiration (Exps. 152 and 154). In the second place, transpiration decreases in darkness and increases in light, though the fact that the temperature may be considerably lower in the dark room than in the open air is in part responsible for the lower rate of transpiration in the former situation.

Bearing in mind that the stomata of many plants close

in darkness and open in light, that they may close in dry air or when the plant is dry at the root, and that they open under circumstances in which evaporation is reduced, we may conclude that, though transpiration is set up as the result of evaporation of water-vapour from the air-spaces surrounding the cells of the leaf, though it is hastened by conditions which favour evaporation, and checked by conditions which hinder evaporation, the process of transpiration is not a mere evaporation. For, on the one hand, the living cells resist desiccation automatically; as they become drier, so the contents of their sap concentrate; the osmotic pressure increases, and water is held more strongly. On the other hand, by the opening and closing of its stomata, the plant regulates to some extent the rate of loss of water. As the guard-cells lose water, they decrease in size, change in shape, and come closer together, so that the stoma between them is made smaller. If water is supplied plentifully, e.g. by wetting a leaf, the guard-cells become turgid, the stomata open, and the rate of transpiration tends to increase. Whether the prime purpose of the movement of stomata is to secure this end, or whether this is a secondary matter, we cannot decide. It may be that, as the plant loses water, it becomes important for it to shut down all its activities, and that the closure of the stomata serves to secure this end by reducing the rate of gaseous exchange between the plant and the air.

CHAPTER X.

THE passage of water from root to leaves: the channels followed by the transpiration current: water-conducting wood and skeletal wood. The causes of the ascent of water. Phenomena connected with the absorption of water: root pressure: bleeding: excretion of water: water-pores (hydathodes).

OUR previous studies have taught us that of the water absorbed by the root hair-cells, some passes to neighbouring living cells and, absorbed osmotically by them, contributes materially to the increase of weight manifested by the growing plant. The water used in this way is, however, but a small part of that taken up by the root. By far the larger part of the water absorbed travels, as we know, from the root to the leaves, whence it passes into the air in the form of water-vapour. The stream of water which passes through the plant we may call the transpiration current, and we require to learn what are the channels along which this current travels in its passage from the root to the leaves. That the transpiration current passes along definite channels we demonstrate in the following way:—

155. Dig up one or two well-grown seedling bean or pea plants, wash their root-systems free from earth, cut across the main roots, place the plants with their roots in water in which is dissolved some dye, e.g. eosin or red ink, and leave them exposed to the light till the colour of the dye is visible in the "veins" of the leaves and also, if the plants are in bloom, in those of the flowers. Cut the root and shoot of one specimen transversely into a series of pieces, examine the cut surfaces by means of a lens, and determine that the dye has travelled up

the stem along definite tracts. Note that the position of the stained areas is not the same in the root as in the shoot.

156.* Mount transverse sections of the stem in water or dilute glycerine, and examine them microscopically. Note that the tracts along which the dye has passed contain what appear to be largish spaces, each surrounded by a fairly thick and stained wall. Observe that the dye which stains the walls of the relatively wide water-channels is absent from the cellular tissues of the stem. By means of similar cross-sections, follow the course of the dye from the shoot, through the leaf-stalks to the " veins " of the leaf.

157. Cut the root and shoot of the second plant lengthwise, proceeding cautiously from below upward, and trace the longitudinal course of the dye. By the aid of the microscope, observe, in thin longitudinal sections, that, whereas the tissues of the part outside the stained areas are made up of minute cells, those in the stained areas contain long and wide tubes, the walls of which are curiously thickened in an annular, spiral or other regular manner. These dead elements are, as the dye indicates, the channels (vessels) along which the water passes to the leaves. On tracing them towards the leaves, the vessels are found to connect with similar, smaller groups which run in definite courses through the leaf-stalks, and are distributed in the " veins " throughout the leaf. Some idea of the length of individual vessels may be obtained in the following way :—

158. To ordinary leaf-gelatine add so much water that it remains liquid at about 35° C., but sets to a solid mass at room temperature. Colour the fluid gelatine with eosin, and pour it into a tall jar, *e.g.* a measuring vessel, stand the latter in a large beaker or saucepan containing water which is kept at a temperature of about 40° C. by means of a bunsen flame. Plunge the cut end of a leafy branch below the surface of the gelatine, and leave it for an hour. Remove the vessel containing the branch from the beaker of hot water, plunge it under a tap, and allow a stream of cold water to fall on it so as to cause the gelatine to set. When the branch is cooled, take it out of the gelatine, cut it longitudinally, and determine the length of the stem along which the gelatine may be traced.   Inasmuch as the gelatine is free to pass into the open ends of the vessels exposed on the cut surface, but cannot pass across the end walls of the vessels, we obtain a rough measure of the length of the vessels in the plant on which the experiment was made.

By more accurate methods it has been demonstrated that the length of vessels varies very considerably in different plants.   Thus, in the oak, they have been estimated to be 2 metres or more long, in species of fig (Ficus) they range from 10 to 66 centimetres, whereas the much shorter conducting elements (tracheids) of the Scotch Fir (Pinus sylvestris) are only about 4 mm. in length.

159. Whilst the previous experiment is going on, cut off two similar branches, *e.g.* of laurel, stand one with its cut end in water, and plunge the cut end of the other beneath the surface of melted coco-butter.   After about quarter of an hour, remove the branch, cool its cut end, make a fresh cut about a quarter of an inch from the exposed surface, stand the branch with its cut end in water in bright light, and observe that, owing to the vessel having been blocked by the coco-butter, the leaves wither far sooner than those of the branch which was placed at once in water.

The conclusion that water passes through the *cavities* of the vessels may be verified thus :—

160. Set up a well-grown branch of sunflower (Helianthus annuus) or Jerusalem artichoke (H. tuberosus) in a potometer (Exp. 151), and determine its rate of transpiration.   Fix the middle of the branch in a vice and screw the latter so tightly as to compress the stem and hence its vessels.   Observe that the rate of transpiration falls off very markedly.   Release the stem from the vice and ascertain that the rate of transpiration increases.

The microscopic examination of the cross-section of the bean stem of Exp. 155 showed us that the vessels are distributed in isolated groups.

161. That this is not their distribution in older stems we recognise by observing with the aid of a lens the cut ends of two-, three- and more year-old branches of woody plants, *e.g.* trees such as lime, lilac, etc.   In these plants, the wood forms the greater part of the stem, and the vessels, which make up the larger part of the wood, are seen to be distributed uniformly throughout the woody tissues.

The process by which the wood is increased from isolated strands to a solid cylinder may be studied in text-books of plant-anatomy (Bibliography, 5, 6), but, by comparing the cut surfaces of one-, two-, and three-year-old branches with one another, we observe that each year a new ring of wood is formed on the outer side of the old ring. We note further, that there are fairly sharp lines of demarcation —the annual rings—between each year's wood and that of preceding and successive years. With a good lens, or by microscopic examination of sections, observe that the appearance of these concentric annual rings is due to the fact that the vessels which are the last to be formed in late summer of one year are thicker-walled and have smaller cavities than those which are the first to be formed in the spring of the ensuing year. We conclude that in our native perennial plants, which possess well-marked annual rings, it is possible to determine the age of the tree by inspection of the surface exposed by a cross cut. Though we cannot now study the question as to the cause of these marked differences between the wood formed in spring and that formed in late summer, we may remark that it is only in regions like our own, in which vegetation is subject to a distinct seasonal check, that the wood shows well-marked annual rings.

Our main object in referring to this matter is to ask our-selves the question, whether all the vessels of the wood are capable of conducting water, or whether this task is dis-charged by certain vessels only? The fact that there are to be seen hollow trees, such as willows, which have lost much of their older wood—that nearer the centre of the trunk— seems to indicate, that at all events in such plants, the young wood suffices for water-conduction. The first step towards an answer to this question we make as follows :—

162. Obtain four, similar, leafy branches of some tree, e.g. oak, elm, etc., and stand them with their cut ends in water coloured with eosin or red ink. Having formed a clear idea of the extent of the wood by inspection of the cut surfaces, remove a ring of bark from the stem of one specimen (No. 1), avoiding cutting into the young wood. Slip a rubber band around the stem of No. 2, and, using the edge of the rubber band as a guide, make a circular cut with a sharp knife into the bark and young wood. In No. 3 make a similar circular cut, but so deep that only the oldest, central heart-wood remains intact. Make no incision into No. 4. Ascertain, by the potometer method (Exp. 151) or by observing the rates of withering of the leaves, what parts of the wood are concerned specially with water-conduction.

By means of such experiments on different plants we demonstrate (1) that the parts of the stem external to the wood are not concerned with the transmission of the water-current along the stem; (2) that the young wood is specially concerned with water-conduction; (3) that the older wood of such trees as oak is incapable of conducting water, though in others, e.g. lime, it aids the young wood in this work. In those trees, the old wood of which has ceased to serve the purpose of water-conduction, it may be noted that marked differences of colour, etc., exist between old and young wood; in these cases, the difference between younger splint wood (alburnum) and older heart-wood (duramen) is made visible. It will be evident that though the heart-wood of a given tree has ceased to conduct water, it nevertheless may play an important part in acting as an internal skeleton to the trunk, and thus giving it the power of resisting the shearing action of wind and providing a support for the ever-increasing crown of branches which the tree puts forth. Though we are not now making a thorough study of the structure of wood, we must satisfy ourselves that the woody tissue does not consist exclusively of vessels, but contains other elements, some of which may be shown by appropriate tests (Bibliography, 2) to be unlike the wood, not only in size, but also in the fact that they contain living protoplasm and also reserve-substances, such as starch.

We are now able to picture to ourselves the water-conducting system of the plant. Below, in the soil, are the root-hairs, the organs for the absorption of water. The inner walls of the root-hair cells adjoin those of the cortical cells which absorb, according to their osmotic capacity, water from the root-hair cells and become turgid. The innermost cortical cells are in contact with others of the deeper-lying tissues, and these, in turn, abut on the vessels.   In its young stage, a vessel, which is formed from a longitudinal row of cells, is a living element, and, like all living cells, contains osmotic substances.   Hence water passes by osmosis into the developing vessel.   The constituent cells which are forming the vessel increase in length, their cross walls disappear, their longitudinal walls thicken, become lignified (woody), and are rendered rigid by the bars or rings or spiral bands of woody substance formed, as the last act of the protoplasm, on their inner surfaces.   Thus, the completely developed vessel is full of water.   Each such vessel of the root connects above with a similar older vessel, the cavities of the two vessels being separated only by a thin membrane which constitutes the end walls of the two vessels.   Across this membrane, water and dissolved substances pass readily.   If we think of the plant as it grows from the seedling stage, we can picture new vessels filled with water being added in root and stem by the development of rows of cells some distance behind the growing points; and hence we can imagine continuous columns of water in each of the longitudinal series of vessels which extend from the veins of the leaf, through the stem, to the root.   Now, it has been shown that a continuous column of water, such as we have imagined to exist in the longitudinal series of vessels in the plant, has very remarkable properties.   Even though it is subjected to a considerable pull, the column does not break.   We know that, in the plant, water is evaporating from the inter-cellular spaces of the leaves into the air, and that this process sets up a chain of events, resulting in the withdrawal of water from the vacuoles of the green cells. Hence the osmotic pressure of these cells is increased, and they withdraw water from the colourless cells with which they are in contact.   The colourless cells, abutting on the vascular elements in the veins of the leaf, absorb water from these elements.   But as we have pictured the conditions, these vascular elements contain the tops of continuous water-columns, which extend through the stem to the root. As water is lost above, one of two things must happen, either the water-column must break, or the column as a whole must be hauled up. It can be shown experimentally that such a column does not break, even when subjected to a very considerable pull. Hence we may infer that it is hauled up bodily. Thus it may be in this manner that the ascent of water is effected even in great trees, such as the Eucalyptus of Australia, or the Sequoia of Western America, which reach a height of about 300 feet.

It should be noted that we do not offer proof that this is actually the mode by which the ascent of water is effected; we may, however, adopt it as an hypothesis to account for the fact—which is one of the most remarkable in the whole range of plant-physiology—that a tree is able to haul up large quantities of water to a very great height. Although it must not be regarded as a model illustrating the mechanism of the lift of water which takes place in transpiration, it is none the less instructive to make an apparatus which shows that, when water is evaporating from a surface connected with a water-column, a considerable " pull " is exerted. For this purpose we proceed as follows :—

163. Soak a piece of parchment membrane in water and tie it tightly over the wide end of a thistle funnel. Fill the funnel *completely* with boiled water, and, closing the narrow end with the finger, stand it with the wide end uppermost in a dish containing mercury. Place a scale behind the apparatus and record from day to day the height to which the mercury is lifted in the tube as a result of evaporation of water from the surface of the parchment.

164. A better apparatus, but one that requires a little skill to construct, is made by causing plaster of Paris to set so as to form a layer closing the mouth of the funnel. The apparatus is filled with water, and in this, as in the previous experiment, it is important that the water used should be rendered by boiling as free from air as possible, since the air contained in unboiled water, being liberated as evaporation proceeds, accumulates beneath the surface of the plaster and tends to break the continuity between ·

the water in the plaster and that in the tube. That the transpiring plant exerts a similar pull may be demonstrated in a similar way as follows :—

165. Cut a branch from an actively-transpiring plant and stand it for some hours in water. Fit it, by means of a split cork, into a vertical tube filled with water. This may be done by plunging the tube and the cut end of the branch in a pail of water and placing the cork in position. Raise the tube, fix it in the clamp of a retort-stand, wipe the cork dry, and make the joints air-tight by means of a layer of wax. Whilst the lower end of the tube is closed with the finger place the apparatus so that it just dips into a dish containing mercury. The rise of mercury in the tube is then recorded as in the previous experiment.

Another experiment may now be performed to demonstrate the fact that the activity of the roots in absorbing water may, especially at certain seasons of the year, set up a considerable pressure in the plant, which is described as *root-pressure*. Not all plants exhibit this phenomenon, and hence root-pressure is not to be regarded as an agent effecting the ascent of water. The experiment may be tried on any convenient vigorous plant : Sparmannia africana, or species of Phaseolus or Fuchsia are suitable for the purpose.

166. Give the pot-plant selected for the experiment a thorough watering. Cut back the stem to within a few inches of the soil. Slip a piece of rubber tubing over the cut end and wire it securely ; pour in a little water and fix, by means of wire, a long glass tube of fairly narrow bore in the other end of the rubber tubing. Place a scale behind the tube and record at intervals the height of the water in the tube. In the case of trees and shrubs in the open, experiments to demonstrate root-pressure succeed best in spring before the leaves are fully expanded. The phenomenon of root-pressure may be connected with others occurring in nature. For instance, under certain circumstances, vines, especially when pruned late and with the wounds imperfectly healed, exhibit a " bleeding " process. In the spring, sap containing various substances in solution exudes from the cut ends of the shoots.

The quantity of fluid which may escape by bleeding is often considerable, *e.g.* from birch, sugar maple, palms, etc. The liquid which escapes from the injured flower-stalks of certain palms is so rich in sugar that it is used in Ceylon and Java for the manufacture of alcoholic liquors. In some cases, this exudation or "bleeding" may be due to root-pressure, but, in others, it is the result of pressure set up locally, often as the result of an injury to the stem or other parts. Thus, in Java, the natives, having beaten the inflorescences of palms with wooden hammers, cut off the flower-stalks, and fix vessels to catch the liquid which exudes from the wounds.

Connected with phenomena of bleeding are those exhibited by certain plants, *e.g.* oat, fuchsia, saxifrage, etc., which show frequently in the morning drops of water, often mistaken for dew, glistening on the edges or surfaces of the leaves (p. 2, Chapter I.). In some tropical plants, *e.g.* species of Colocasia, this phenomenon of *guttation* is so marked that at intervals of a few minutes drops of water drip like rain from the tips of the leaves. In these cases, it is found that the leaves possess special organs, often formed by the modification of stomata, known as hydathodes.[1] In various plants, notably in species of saxifrage, the water exuding from the hydathodes contains carbonate of lime in solution, which, on the evaporation of the water, is deposited as a white fringe on the edges of the leaves. Certain water-plants, *e.g.* Ranunculus aquatilis, Callitriche verna, etc., possess very conspicuous hydathodes at the tips of the leaves, which either open directly to the exterior or are covered over only by the cuticle. It is probable that, in water-plants, these organs serve as means of increasing the rate at which water is discharged by the leaves, and so of augmenting the supply of dissolved

---

[1] The phenomenon of "guttation" is most easily observed when the plant is growing under conditions which, whilst favouring absorption of water by the roots, hinder rapid transpiration, *e.g.* in the early morning in a warm, moist greenhouse, or, if the plant is kept under a bell-glass, in a warm room. The excretion of liquid water may be observed in any of the following plants:—Primula sinensis, Tropaeolum majus, Aconitum napellus, Delphinium sp., Fuchsia sp., Helleborus niger, Vicia sepium, Triticum vulgare, Avena sativa, Zea Mays. (See list of plants, Appendix.)

mineral salts. Owing to the fact that many of these plants have submerged leaves, the transpiration of water-vapour does not occur; indeed, in some cases, stomata are not present, and hence the movement through the plant of water containing dissolved salts is likely to be very slow. The explanation of the significance of hydathodes in the case of terrestrial plants may be as follows :—During the night, the roots continue to absorb water, but transpiration, as we have already indicated, is considerably reduced. The water-courses, and the tissues of the plant generally, become filled with water, and the increasing pressure is relieved by the exit of water through the " water-pores " or hydathodes, which may thus be likened to safety-valves.

In concluding our study of the relation of plants to water, we note that, beside effecting the transport of dissolved mineral substances to the leaves, transpiration serves also to prevent the undue heating of leaves exposed to bright sunshine. Just as perspiration, evaporating from our bodies, lowers the surface-temperature, so transpiration lowers that of the leaf. Finally, the fact that, during bleeding, sugars and other organic substances may be exuded from the vessels, may be taken to indicate that, not only are the vessels the conduits along which water and salts (crude sap) are transported, but also that they may serve, upon occasion, for the transport of " elaborated sap," that is, of plastic food-materials. In spring, at all events, when the unfolding leaves are making a heavy demand on other parts of the plant for food-substances, it appears that the demand is met by the discharge of sugar and other plastic food-substances in solution into the vessels through which these substances are transported rapidly, in the transpiration-stream, to the leaves.

Before leaving this subject, we demonstrate that transpiration—that is, the regulated process of evaporation of water—is not confined to the leaves. As the results of Exp. 138 indicate, the shoot-axis may also transpire.

167.* Microscopic examination of sections across a young shoot shows that stomata are present in the epidermis, and we demonstrate, by means of cobalt chloride paper applied to a portion of a young stem, that the stomata serve as channels for the escape of water-vapour.

168. In older stems, *e.g.* two- or more year-old branches of trees or woody shrubs (elder, etc.), the stomata are replaced by less regular openings, called lenticels, which, as we may show by the cobalt chloride method, serve for the exit of water-vapour. The lenticels in the elder, etc., may be recognised by the naked eye as small elongated areas, differing in colour from the rest of the surface of the stem. That they serve as channels of communication between the external air and that in the intercellular spaces of the plant we demonstrate thus :—

169. In summer, cut off a two- or three-year-old shoot of elder or horse-chestnut. Remove from it a piece bare of leaves and several inches in length. Attach to one end a short length of rubber tubing, turn the tubing on itself, and bind it tightly to that on the stem so as to close it completely. To the other end of the piece of stem, attach a thick indiarubber tube (pressure tubing) about a foot long. Insert into the free end of the pressure tubing a stout glass tube several feet in length. Make the joints good; lay the stem in a vessel of water; add water through the glass tube, and, by squeezing the rubber, drive out the air contained in the pressure tubing. Clamp the glass tube vertically in a retort stand. Pour mercury into the glass tube by means of a funnel. Note that, as the height of the mercury column increases, streams of air-bubbles escape from the lenticels. Compare with Exp. 146, and apply a similar explanation.

170.* Microscopic examination of a cross-section of the superficial tissues of the old stem of elder, etc., sup-plies the explanation of the behaviour of the lenticels in allowing the escape of gases. Whereas the outer layers of the shoot are made up generally of brick-shaped cells, fitted closely together, the lenticels consist of rounded cells, between which are definite intercellular spaces. The brick-shaped cells are empty, dead, and have thick, corky walls. A section through a small piece of bottle-cork shows similar layers of brick-shaped cells. Now cork is used domestically to prevent the escape of water and gases. Man, in fact, puts cork to a use similar to that to which it is put by the plant.   The layers of cork which occur in the older shoots of woody plants serve to prevent loss of water by evaporation from the thin-walled, delicate, underlying tissues.   But a continuous jacket of cork about a branch would not only mean prevention or reduction of loss of water by evaporation, it would also mean prevention or reduction of all gaseous exchange. The tissues of the stem would be deprived of oxygen, and, consequently, their growth and activity would be checked. The lenticels, like the stomatal openings of the leaf and of the young stem, serve as aerating channels, and hence also as channels for the escape of water-vapour.   We thus discover that the plant, by developing a "water-proof" layer of cuticle in the outer walls of its epidermal cells in the young parts (leaves, young shoot), and by forming sheets of cork in or beneath the epidermis of its older stems, contrives to meet two needs ; (1) the prevention or reduction of indiscriminate evaporation, and hence the regulation of transpiration ; and (2) the provision of aerating channels whereby gaseous exchange between the air and the tissues of the plant is maintained.   If oxygen is being used by any of these tissues, it is derived from that in the air of the intercellular spaces.   The oxygen pressure of that air falls.   When it falls below that of the external air, diffusion of oxygen molecules from the external air to that of the intercellular spaces is set up, and thus the cells abutting on the intercellular spaces receive a constant supply of oxygen.   Similarly, any gases produced by the cells of the tissues of the plant may escape by diffusion into the outer air.

We have a means of verifying the accuracy of the twofold significance of lenticels in relation with transpiration and with gaseous exchange.   In winter, the plant is, compared with its summer state, at rest.   The leaves fall, and the buds remain till spring enclosed within hard scale-leaves.   In such circumstances, the plant has need for but little oxygen, and hence is under no necessity of keeping its aerating channels open.   Moreover, since, in winter, the soil-temperature falls so low that the root-system is able to absorb but little water, the existence of open passages through which water-vapour may readily escape would be a source of danger to the plant, leading to its death by drought. Therefore, if our interpretation of the twofold significance of stomata and lenticels is correct, we should expect to find that the plant takes steps to put them out of action during winter. With respect to the leaves, we know that in most plants this is so : they are, in deciduous trees, put out of action altogether by being cast off from the branches. With respect to herbaceous perennials, the prevention of loss of water is effected by the most drastic means—the whole plant dies down to the level of the ground, and winters underground.

171. That the lenticels also are put out of action in winter we demonstrate by repeating, at that season, Exp. 169. We find then that, in spite of considerable pressure, no air escapes from the lenticels. They are closed.

172.* By means of sections we prove that, when the plant is preparing for winter, it produces, in its older stems, new sheets of cork below those already in existence, and instead of forming lenticels of loose masses of cells as it does in spring, it produces a continuous corky layer beneath the lenticels as well as elsewhere. Thus transpiration is slowed down and gaseous exchange is reduced.

It is indeed remarkable that the fall of the leaf is engineered in many plants in precisely the same manner. Thus, in the horse-chestnut in late autumn, cells at the base of the leaf-stalk begin to divide, forming an active layer of brick-shaped cells. These cells divide again to form tiers of cells, which become corky. The corky cells form a layer around the base of the leaf-stalk, and moreover the water-channels running through the stalk to the blade become blocked. The leaf, thus shut off from fresh supplies of water, gradually withers, dries, and, at a breath of wind, comes rustling to the ground. This is facilitated by the formation of an " absciss layer " on the leaf-stalk side of the corky band. It consists of a few layers of delicate cells formed by rapid division of a row of cells shortly before the leaf falls. The corky layer serves the purpose of healing the wound, preventing loss of water, ingress of spores of disease-producing fungi, etc. Each leaf that falls is cut off by a surgical operation performed by the plant, and the wound produced is healed in advance.   Similarly, as a consequence of the formation of cork in the stem, all the tissues external to the cork are cut off from supplies of water, and hence dry, shrivel and crack.   Trees which produce cork in the surface layers of their stems have thin, smooth bark; those which produce it in their deeper layers have thick, corrugated bark.   According to the exact mode of development of the cork, the bark is seen to be in sheets as in birch and beech, in lozenge-shaped masses as in elm and oak, and so on.

When growth resumes in spring, not only are new leaves formed, but, in many trees, by the growth of the deeper tissues of the stem, the bark is split and torn, and may be cast off.   Many plants utilise the tissues which constitute the bark for excretory purposes, that is, for getting rid of waste-substances.   Thus it is that many and varied kinds of materials, some of which are of great value in commerce and medicine, are obtained from the bark of trees (quinine, tannin, hazeline, etc.).

Using the nomenclature of p. 149, we may say that, in winter, a deciduous tree, leafless and girt in cork, is a xerophyte; in summer, crowned with delicate leaves and with open lenticels, it is a hygrophyte.   Such plants, which change periodically from the xerophytic to the hygrophytic state are called tropophytes: and to the fact that, as soil-temperature falls, absorption of water decreases below the amount necessary to make good the loss due to transpiration is to be ascribed the wonderful series of changes which mark the preparation of the plant for passing the winter.   The subject is an alluring one, but we cannot follow it further now.

173. The changes of some common tree should be followed throughout the year and records of the dates of bursting into leaf, of closing of the lenticels, of cork formation, etc., should be made; and the records compared with meteorological records, particularly with those of soil temperatures.

CHAPTER XI.

THE origin of the carbon compounds contained in plants. The raw materials from which the plant constructs these compounds. The part played by chlorophyll grains (chloroplasts) in the manufacturing process: the energy by which the process is carried on. The passage of carbohydrates from the leaves to other parts of the plant. The synthesis of organic nitrogen compounds by the plant.

BEFORE we embark on the last stage of our enquiry into the mode of nutrition of plants, we will review briefly the conclusions to which our experiments have led us.

Sand- and water-cultures have served to demonstrate that certain mineral substances are essential to plants, and also, that all the elements required by the plant for its nutrition are, with the exception of carbon, derived from the soil in the form of water and mineral salts. Of the water and mineral salts absorbed by the root, a certain amount passes by osmosis from cell to cell, and thus satisfies local needs, but the great bulk travels by way of the transpiration current to the leaves. There the water is transpired and the mineral substances are left behind. Thus the green cells of the leaves receive constant supplies of mineral substances. Since these mineral salts do not accumulate indefinitely in the leaves, and since also the elements which they contain become constituents of complex organic bodies, such as proteins, it is evident that the mineral salts, in serving for nutrition, undergo chemical change. Therefore, they are to be regarded, not as food-substances, but as raw materials used by the plant in the manufacture of food-substances. Similar considerations convince us that the carbon-containing organic substances occurring in the plant arise as the result of processes of manufacture carried on by the plant, and that the carbon which finds a place in these organic compounds is derived from external sources.

It is our object in this chapter to discover what we can of the nature of this manufacturing process, and we commence by enquiring into the source whence is derived the raw carbon-containing material used in the manufacture of organic compounds.

The problem is simple. We know that a plant may grow and that the amount of its carbon-compounds may increase, although there may be no compounds of carbon present in the soil. Therefore, the air is the source whence the plant derives its carbon. It must, however, be admitted that this does not appear, at first sight, to be very probable. For, on the one hand, the amount of combined carbon contained in a plant—a tree, for example—is very great; on the other hand, the only carbon-containing constituent of pure air, carbon dioxide, makes up but a minute fraction ($\cdot$03 %) of the total volume of the atmosphere. It is doubtless these facts which made botanists, down to comparatively recent times, slow to accept the result of the earlier experiments which pointed to the air as the source of carbon to the plant. They preferred to adhere to the old " humus theory," which taught—seemingly on good grounds—that the organic compounds of carbon in the soil are the sources whence the green plant obtains its supplies of this element. They knew that plants flourish when provided with a plentiful supply of organic carbon-compounds, when, for instance, the land is richly manured; they noticed that soils deficient in these substances support but poor crops, and it seemed to them a breach of common-sense to be asked to believe that the air, with only some three parts in 10,000 of carbon-dioxide, could provide the large quantities of carbon which are contained in plants. A more thorough consideration of the facts shows us that, though the *percentage* of carbon dioxide in the atmosphere is but small, the actual quantity contained in that vast ocean of air is enormous. Moreover, this quantity is augmented constantly. Every fire that burns, every plant or animal that respires, and every organic substance which undergoes decay,

K.P.   M

produce carbon dioxide, and discharge it into the atmosphere.

But, after all, experiment is better than argument, and we proceed, therefore, to put the matter to experimental proof.

If the carbon dioxide contained in the air is the raw material used by the plant in the manufacture of organic compounds, then we shall expect to find that an adult plant supplied with air, from which the carbon dioxide has been removed, ceases to manufacture organic carbon-compounds. Since, as our studies of germination have shown, organic carbon-compounds, such as sugar, serve for the nutrition of the plant, it follows that, if there are no stores of organic carbon-compounds present, by preventing the plant from manufacturing these compounds, we reduce it to a state of starvation. To carry out the experiment, we proceed as follows :—

174. Take two approximately equal, actively-growing young sunflower or dwarf bean plants raised in small pots, label them A and B, cut off a piece about half an inch square of one of the older leaves of each plant. Having watered the plants, place them in the dark. At daily intervals, cut off similar samples. Keep the samples separate, and immediately after they are cut off, plunge them for a minute into boiling water to kill them. Place the pieces in wide specimen tubes containing methylated spirit. Label each tube with the letter A or B and the date on which the piece it contains was cut off. Cork the tubes and expose them to bright sunshine. After a piece has been in a tube for 24-48 hours, take it out, observe that it is brittle and colourless, and rinse it with water. Place each sample separately in a white saucer, and pour over it a solution of potassium-iodide iodine, to which has been added a strong solution of chloral hydrate (Appendix A). Observe that, as indicated by the iodine reaction, when the plants are maintained in the dark, the starch which is contained in their leaves gradually decreases in amount until, after one or more days, it disappears altogether. When this has happened, the plants are ready for use. They should, however, be kept in darkness till the apparatus now to be described has been prepared.

175. Having obtained two large bell-jars—(the cloches

(Appendix A) used in "French Gardening" serve admirably), take one of them to a carpenter and get him to make the following :—A circular, wooden base-board about an inch thick and of a diameter about two inches larger than that of the rim of the bell-jar. The base-board must be of one piece, and should be varnished. Three, 3-inch cubes of wood to serve to raise the base-board above the level of the table or ground. Instruct the carpenter to cut on the base-board a circular groove, about $\frac{3}{8}$ inch in depth and $\frac{1}{2}$ inch wide, of such diameter that the rim of the bell-jar fits comfortably into it, and to bore a round hole 1-$1\frac{1}{2}$ inches diameter, at a distance from the groove about equal to one-third of the diameter of the circle described by it. The edge should be circular and smooth since a rubber cork is to be inserted in the hole.

Bring the apparatus into the laboratory, and select a glass tube of such a size and shape that, when it is in position under the bell-jar, its wide end is an inch or more from the top of the latter, and its narrow end projects below the bottom of the base-board, but clears the table by about 2 inches. A cylindrical separating funnel with a narrow stem serves the purpose (Appendix A). Push a small plug of cotton wool to the bottom of the wide end of the tube, and fill the latter with soda-lime—a substance which absorbs carbon dioxide. Insert the stem of the absorption tube into a rubber cork which is large enough to fit tightly into the hole of the base-board. Fix the rubber cork in position ; make a good joint ; attach to the end of the stem which projects beneath the board a piece of rubber tubing, and close the latter with a clamp. Set up the board in a well-lit place—preferably in a sheltered position in the open-air but screened from direct sunlight. Bring quickly one of the plants (A) from the dark room. Make *rapid* measurements of the height of its stem and the size of its smallest leaves (this need only take a minute or so). Stand the plant on the board, place the bell-jar in position, and cover the bell-jar with a double fold of black Italian cloth in order to protect the plant temporarily from the light. Having melted enough wax-mixture (Appendix A) for the purpose, pour it into the groove so as to seal the bell-jar (or soft putty may be used). Bring the second plant from the dark, measure its stem and youngest leaves. Cover it loosely with a bell-jar. This second apparatus serves as a control. Remove the black cloth from the first bell-jar, unclamp the rubber tube so that air may now enter the bell-jar through the absorption tube. We thus have the two plants under fairly similar conditions, except that one receives air deprived of carbon dioxide, the other ordinary air. When the plant which is exposed to ordinary air shows a definite increase both in height of stem and size of the younger leaves, estimate and record the increase, noting also the number and size of any leaves it may have unfolded since it was put under the bell-jar. Remove the wax from the edge of the other bell-jar, take out the plant, cut off one of its older leaves, and, having put the leaf in a dark place, determine that the plant has grown but little or not at all, and that it has developed no new leaves. Make these observations as quickly as possible. As soon as they are made, cut off a leaf from the control plant. Before examining the cut leaves, place the plants in the dark. Make a deep cut in one of the two isolated leaves in order that it may be distinguished from the other, and test the leaves for starch by the iodine method. Observe that the leaf taken from the plant which was exposed to ordinary air gives a well-marked, blue starch reaction, and that the leaf of the plant which was deprived of carbon dioxide gives no starch reaction. Bring the plants from the dark room into bright sunlight. At intervals of five minutes, take samples of the leaves of the plant which was under the sealed bell-jar, and, by means of the iodine test, demonstrate that, within ten minutes or so of the time at which plant A was exposed to the light, starch makes its appearance in the leaves, and that the amount of starch—as judged by the depth of blue colour—increases as time goes on. Note, further, that the plant, the development of which was checked completely when it was in air devoid of carbon dioxide, may begin again, in the course of a day or two, to increase in size, and to unfold new leaves.

We learn from this series of experiments that :—

(1) The leaves of a green plant, grown under ordinary conditions, contain starch.

(2) Starch disappears from the leaves of a plant kept in darkness.

(3) Starch reappears very quickly and in increasing quantities when the plant is brought again into the light and exposed to normal conditions.

(4) If a plant is deprived of carbon dioxide, no starch appears in the leaves, even though it is exposed to a good light.

(5) In the absence of carbon dioxide, the growth and development of the plant are checked.

From the results of our experiments on the germination of seedlings, we know that reserve starch stored in the seed serves to provide the seedling with carbohydrate food-material. Hence it is probable that the starch which makes its appearance in the leaves of plants grown in the light in ordinary air serves also to provide for the nutrition of the plant. Since, moreover, starch does not appear in the leaves of plants grown in air deprived of carbon-dioxide, it is probable also that carbon-dioxide serves as the raw material from which carbohydrate, which appears in the leaf as starch, is manufactured. If this is so, the leaves—and other green parts of a plant— are the seat of a manufacturing process which is of fundamental importance not only to the plant itself, but also indirectly to animals, all of which obtain their food ultimately from plants. But since the leaves of plants kept in darkness do not form starch—indeed, as we have seen, this substance actually disappears from them under these conditions—it follows that, if starch is manufactured by the leaves of plants, light plays an essential part in the process. Now, it is easy to demonstrate that the starch which appears in the leaves of plants exposed to the light is actually manufactured in the leaves, and is not carried thither in the form of sugar from other parts of the plant.

176. For this purpose, take a well-grown pot-plant (bean, clover, etc.) and keep it in the dark till its leaves—as determined by tested samples—contain no starch. Then detach one or more leaves from the plant, place their petioles in distilled water and expose the leaves to the light. After an hour or so, on applying the iodine test (cf. Exp. 174), we discover that starch has made its appearance in the isolated leaves.   Inasmuch as these leaves are isolated from the plant, the starch which appears in them cannot be derived from substances transported to the leaves from other parts.   Therefore, either it must have been manufactured from raw materials or from some substance, such as sugar, already present in the isolated leaves.   Since, however, as Exp. 174 has proved, starch does not appear in leaves which, after they have been depleted of this substance by a sojourn in the dark, are brought into the light and confined in a vessel from which carbon dioxide is excluded, we conclude that the starch which appears in the illuminated, isolated leaves arises as the result of a process of manufacture in which raw materials, of which carbon dioxide is one, are employed.   That this is the only conclusion open to us, we demonstrate further by showing that the dark-kept leaf contains little or no sugar.   Thus :—

177. Take two lots of leaves, one from the dark-kept plant of Exp. 175 and one from a similar plant which has been exposed to a good light for some hours.   Cut them in pieces under water, pound each lot separately in a mortar with a little water, transfer each mass to a beaker, boil, and filter.   Test the filtrate for sugar.   Observe that the dark-kept leaves give no, or at most a slight, sugar reaction, and that the light-exposed leaves give a well-marked sugar reaction.   We thus learn that not only is starch produced when the green parts of leaves are exposed to light, but that sugar also is formed.   Since this sugar and starch can only have been formed from raw materials (carbon dioxide and water), we may speak of their formation as a synthesis or building up, and since light is essential for their formation, we may describe the process as one of *photosynthesis* of carbohydrates.   Inasmuch as it is by this process that the plant obtains its supplies of organic carbon-compounds, the photosynthetic process is termed sometimes that of the *assimilation of carbon*, or of carbohydrates.

The question now presents itself :—Have the green cells alone the power of photosynthesis, or is it possessed also by the colourless cells of plants?   The answer we obtain by the following observations :—

178.* Cut transverse sections through a leaf (beech, privet, sunflower, etc.) gathered during the day-time. Mount the sections in water. Observe that the green colour of the leaf is due to the presence of vast numbers of green granules (chlorophyll grains or chloroplasts), lying in many of the cells. Note that the ordinary cells of the outer layer (epidermis) of both upper and lower sides of the leaf contain no chlorophyll grains, but that chlorophyll grains are present in the guard-cells of the stomata and in the vertical rows of cells (palisade parenchyma) beneath the upper epidermis as well as in the less regular groups of cells (spongy parenchyma) which lie beneath the palisade parenchyma and extend to the lower epidermis. Observe also the large intercellular spaces between the cells of the spongy parenchyma. Run in potassium-iodide, iodine, chloral hydrate solution, and observe that starch grains are present in the chlorophyll-containing cells (including the guard-cells of the stomata) but are absent from the colourless epidermal cells. Note the stained starch grains attached to the chloroplasts.

179. Then obtain one or two leaves of a variegated plant the foliage of which is thin and delicate (variegated maple, Abutilon, etc.). Either photograph the leaf or make an accurate drawing, *e.g.* a tracing of the outlines of its green and colourless parts. Test the leaf for starch by the iodine method and, by comparing the leaf with the original photograph or drawing, determine that the original green areas are stained blue, and hence contain starch, whereas the colourless areas of the leaf are not stained blue—that is, contain no starch. We thus learn that starch occurs only in those cells of the leaf which contain chlorophyll. Hence we infer that the chloroplasts play an essential part in the photosynthesis of carbohydrates.

But, as we learned in Exp. 16, starch may occur in colourless cells, *e.g.* in cotyledons of seeds, in cells of woody tissues, in tubers such as the potato, and so on. Moreover, since organs such as potato tubers are formed and remain underground, the starch which occurs therein does not, like that in leaves, require light for its formation. The well-known fact that green plants cannot live and increase except when growing in the light, combined with the results of these observations, leads us to conclude that the reserve starch of underground organs and other colourless plants is not synthesised from inorganic materials in these storage organs, but is derived from the carbohydrates manufactured by the leaves and other green parts of the plant. On this view, a green leaf not only manufactures carbohydrates to satisfy its own needs, but also supplies carbohydrates to the colourless cells of the plant.

This conception of the leaf as a factory of carbo-hydrates for the plant as a whole, helps us to understand the comparative rapidity with which starch may disappear from the leaves of plants kept in the dark. We can imagine that the starch, accumulated in the leaves, undergoes a change similar to that which the starch, accumulated in a bean cotyledon, undergoes during the germination of the bean, namely, a conversion by the agency of diastase into sugar; and just as the sugar so formed in the cotyledon passes away to nourish, or accumu-late in, the embryo, so that formed from the starch in the green leaf travels *via* the petiole to the shoot, whence it is distributed in all directions, to serve either for the nutrition of growing cells or to form the reserve starch stored in the various tissues.

In order to confirm the truth of this conjecture, we require, firstly, to prove that the green plant possesses the power of converting sugar into starch, and secondly, to obtain evidence that the starch which is found in the green leaf as the result of the photosynthetic process may pass away in some soluble form, such as sugar, from the leaf to other parts of the plant.

180. To demonstrate that a plant is capable of converting sugar into starch we keep a plant, *e.g.* Elodea, in the dark till all starch has disappeared from its leaves. Then, having prepared a 3 % solution of glucose, cut off several of the starch-free leaves and float them on the surface of the sugar solution in saucers and keep them in the dark. After a day or two, apply the iodine test, and observe that starch has made its appearance in the leaves.

181.* By treating sections of the leaves with iodine,

and examining microscopically, we demonstrate that the starch grains occur in close association with the chloroplasts.

Thus we learn (1) that a plant has the power of converting sugar into starch, and (2) that, in the green leaf, this conversion is effected by the chloroplasts. A chloroplast, therefore, appears to have the power of manufacturing starch in two ways : either by photosynthesis or by acting on sugar. The former mode of manufacture occurs only in light, the latter may occur in darkness. The evidence in favour of the view that the starch which accumulates in the leaves of plants exposed to sunlight may be translocated in a soluble form, we obtain thus :—

182. Cut off three similar leaves from a plant, *e.g.* Sparmannia, which has been exposed for some hours to a good light. Put two of the leaves with their leaf-stalks in water in the dark, and, at the same time, place the plant itself in the dark. Test the remaining detached leaf for starch. After twenty-four hours, cut off one or more of the leaves of the dark-kept plant, test them and also one of the detached, dark-kept leaves for starch. Repeat the experiment after forty-eight hours, using the remaining, dark-kept, detached leaf and one or more leaves which have remained attached to the dark-kept plant. Observe that the starch disappears more rapidly from the leaves which remained attached to the plant during its exposure to darkness than from the detached, dark-kept leaves. Similarly, demonstrate by using starch-free leaves of a dark-kept plant that starch may accumulate in a detached leaf (with its petiole in water) brought into the light just as in a similar leaf which is attached to the plant and exposed to light. The latter part of the experiment demonstrates that the activities of the detached leaf have not been impaired by its removal from the plant, and hence the only explanation of the persistence of starch in the detached leaf in the former part of the experiment is that carbohydrate cannot escape from the detached leaf. We conclude, therefore, that, in the normal plant, the carbohydrate produced photosynthetically in the leaf, passes away in the form of sugar to supply the needs of the whole plant.

Now, when we consider the extraordinary rapidity with

which the leaf photosynthesises carbohydrate when exposed
to light (Exp. 175), and also the narrowness of the leaf-
stalk, some of the tissues of which, e.g. vessels and tracheids
of the wood, are concerned with other work, we can understand that the rate at which a leaf manufactures carbohydrate may be greater than the rate at which the product
of this manufacture can pass from the leaf to the other
parts of the plant. This being the case, there is a glut of
carbohydrate in the leaf. Such an accumulation would
undoubtedly—unless it could be got out of the way—im-
pede the action of the machinery involved in the manufacturing work. Thus the idea suggests itself that the
machinery must either cease working, or that the excess of
photosynthesised carbohydrate must be stored temporarily
in the leaf in some form in which it does not interfere with
the working of the machinery. Now, an insoluble, indiffusible, solid substance is less in the way than a soluble,
diffusible substance. Starch belongs to the former type,
and sugar to the latter type. Hence we may suppose that, if sugar is the first product of photosynthesis, and if it is produced by the green leaf faster
than it can be conducted away from the leaf, the excess
may be stored in the form of starch. The chlorophyll
machinery must perforce lie idle during the night, so that,
whereas the manufacturing process can go on only during
the hours of daylight, the work of distribution can be
carried on night and day. The hours of daylight are all
too brief, as it is, for the green leaf to manufacture enough
carbohydrate to satisfy the needs of the growing plant,—to
supply it with food-material and to allow it to accumulate
large stores of starch in reserve organs, seeds, etc. It
would be bad business, therefore, for the photosynthetic
process to be shut down merely because the product could
not be distributed fast enough; it would be good business
if the leaf could store its excess of manufactured material
temporarily, so that the chlorophyll machinery might run
full time. Such considerations suggest to us that the
starch which appears in the leaf is not the direct product of
photosynthesis, but is—as it is in seeds, etc.—a reserve
form, in which the excess of photosynthesised carbohydrate
is stored temporarily.

Let us ascertain if the facts support this hypothesis. In the first place, we have proved already that sugar as well as starch makes its appearance in the leaves when a dark-kept plant is exposed to light (Exp. 181). Hence sugar has at least as much claim as starch to be regarded as the immediate product of photosynthetic activity. In the second place, starch is, as we know, a more complex body than are sugars :—

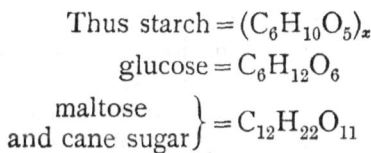

$$\text{Thus starch} = (C_6H_{10}O_5)_x$$
$$\text{glucose} = C_6H_{12}O_6$$
$$\left. \begin{array}{l} \text{maltose} \\ \text{and cane sugar} \end{array} \right\} = C_{12}H_{22}O_{11}$$

and it would seem more likely that the less complex body is produced first.

183. Test for starch and sugar the leaves of such plants as iris, onion, Scilla, which, after having been kept for a few days in the dark, *e.g.* by covering with an inverted flower-pot, have been exposed to the light for some hours. Observe that, whereas sugar is present, the leaves contain little or no starch. We thus learn that, even in leaves in which starch is formed as a result of photo-synthesis, sugars are also present, and that, in other plants, the products of photosynthetic activity are sugars and not starch.

We may conclude, therefore, that some form or other of sugar is the immediate, carbohydrate product of photosynthesis. Of this sugar, some may be used for local nutritive purposes, some may be used locally for respiration, but the bulk is translocated from the leaves to other parts of the plant. Of the sugar thus translocated, part is used immediately for nutritive or respiratory purposes, and part serves one or other of these purposes ultimately, but undergoes, in the meantime, a change into the reserve form, starch. So, too, the excess of sugar produced over that translocated accumulates in the leaf during the day, and is converted into the more convenient storage form of starch, which, however, is changed by diastase back again into sugar as soon as the road is clear for the passage of this substance. Thus the green leaf is, during the day, the seat of carbo-hydrate synthesis, and thus, night and day, there passes from the green cells to all parts of the plant a constant osmotic stream of sugar.   Just as the excess of production over local consumption and over translocation leads to an accumulation of starch in the leaves, so an excess of supply over consumption in any part leads to like storage in the form of starch, and thus it is possible for reserve-organs to assemble large quantities of this substance.

But one point still awaits explanation, viz. :—How is the conversion of sugar into starch effected?

We know that the starch in the leaf makes its appearance in the form of grains attached to, or imbedded in, the chloroplasts.   We know, on the other hand, that reserve-organs, underground tubers, etc., contain no chlorophyll grains.   How then do the cells of such organs contrive to convert sugar into starch?   Now, we have learned already that most plants grown in the dark produce leaves which contain no green colouring matter, and we may ask ourselves the question whether such dark-grown plants are able to store starch in their leaves?   This question is, of course, capable of being solved experimentally.

184. * In order to do this we raise seedlings of Phase-olus or potato in darkness.   Without stopping now to consider the peculiar features presented by the dark-grown, etiolated plants, we proceed at once with our experiment by testing leaves for starch.   Having shown that the leaves contain no starch, cut off several leaves, float them on a 10-20 % glucose-solution in the dark.   At daily intervals, apply the starch-test to one of each of these leaves, and also to others which have remained attached to the plant in the dark.   Having demonstrated that starch is present in the leaves which have floated on sugar-solution and not in the others, cut sections of the former, mount in water, run in iodine solution, and note that the starch-grains are associated each with a yellow-brown stained granule.   Bring the plants from darkness into a good light.   Soon after a distinct greenish colour is visible in the leaves, (a matter of an hour or so) cut sections of a leaf of each plant, observe the chloroplasts, and run in iodine; note that the starch-grains are attached to the chloroplasts, and hence conclude that the chloroplasts of the green, light-grown plant are represented in the colourless, dark-grown plant by colourless granules, which we may call *leucoplasts*. Evidently the leucoplast is a starch-former, and is able to manufacture starch from sugar, and the chloroplast is a more efficient starch-former, for not only is it able to manufacture starch from sugar, but also to manufacture sugar from inorganic materials.

185.* By microscopic examination of young tubers of potato, etc., demonstrate that the starch-storing cells contain leucoplasts, and that the starch-grains are each formed by a leucoplast.

186. Expose potato tubers to a good light, and observe that, in course of time, they become green; that is, their leucoplasts develope, like those of an etiolated leaf exposed to light, the green colouring matter, chlorophyll, which permeates the chloroplast as water permeates a sponge.

We conclude, therefore, that specialised protoplasmic structures, leucoplasts, may occur in the colourless cells of a plant; that they have the function of manufacturing reserve carbohydrate (starch) from plastic carbohydrate (sugar); that the chloroplast is a leucoplast in which green colouring matter, chlorophyll, is developed; and that the chlorophyll confers on the chloroplast the power of manufacturing sugar—in the presence of light—from raw inorganic materials.

187.* It is interesting to note also that the green chloroplasts may themselves undergo a change in colour, as, for example, in the petals and fruits of certain plants, which become brightly-coloured as they ripen. Microscopic examination of various stages shows that such colours are due to small irregular bodies, chromoplasts (or chromatophores), which are derived from chloroplasts. In other cases, however, bright colours, *e.g.* the purple of the copper beech, the colours of the petals of the sweet pea, violet, etc., are, as we determine by the microscopic examination of sections of these objects, and as we have seen in the case of the beetroot, due to pigments dissolved in the cell-sap.

We may express our conclusions as to chloroplasts, leucoplasts and chromoplasts thus :—

plastids	= leucoplasts, chloroplasts and chromoplasts ;
leucoplast and chlorophyll	= chloroplast ;
chloroplast in which chloro-phyll has undergone chemi-cal change and is replaced by a yellow or red pigment	= chromoplast (chromatophore).

We must return now to consider more closely the chemistry of the photosynthesis of sugar, and the part which light plays in this process. So far, all that we know is that, during photosynthesis, carbon dioxide is absorbed and sugar formed. When we consider the formula of a simple sugar (*e.g.* glucose) $C_6H_{12}O_6$, it becomes evident that, in the process of photosynthesis, some substance, which serves as the source of hydrogen must, together with carbon dioxide, take part in the reaction. The only likely source of the hydrogen which enters into the sugar molecule is water. But if we assume that the hydrogen of the photosynthesised sugar is derived from water, then, for carbon dioxide and water together to produce sugar, there must first be a deoxidation or reduction of one or other of these substances. Assuming this, we may symbolise the reaction thus :—$6CO_2 + 6H_2O = C_6H_{12}O_6 + 6O_2$.

If, therefore, our assumptions are correct, we shall expect to find that, whenever photosynthesis goes on, oxygen is liberated. In order to ascertain whether this is the case, we proceed as follows :—

188. Take several green shoots of some common water-plant, *e.g.* Elodea canadensis, tie them loosely together, put them, with their cut ends upwards, inside a large inverted funnel, the mouth of which just fits into a large beaker full of water. Fill a test tube with water, and closing it with the thumb, invert it over the stem of the funnel, which projects upward nearly to the level of the water in the beaker. Thus, any gas given off by the submerged shoots passes up through the water and collects in the test tube. Bring the apparatus into a bright light, and observe that, at once, bubbles of gas are given off from the shoots and rise in the test tube.   Allow the experiment to go on until the test tube is nearly full of gas. This may take several days.   Having prepared a glowing splinter, *e.g.* by lighting a match and blowing it out, lift the tube from the water, plunge into it the glowing splinter and observe that the latter bursts into a bright flame; whence we infer that the gas in the tube consists largely of oxygen.   By more exact methods it may be demonstrated that, as is required by our formula, the volume of carbon dioxide absorbed during photosynthesis is equal to the volume of oxygen evolved.   We thus conclude that a step in this process consists in the reduction of carbon dioxide and water, and we may picture this step thus :—

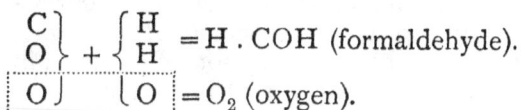

$$\left.\begin{array}{c} C \\ O \\ O \end{array}\right\} + \left\{\begin{array}{c} H \\ H \\ O \end{array}\right. \begin{array}{l} = H \cdot COH \text{ (formaldehyde)}. \\ \\ = O_2 \text{ (oxygen)}. \end{array}$$

On various grounds it is believed that this substance H . COH (formaldehyde) is actually the first product of photosynthesis, and that its molecules condense to form sugar thus :—6H . COH = $C_6H_{12}O_6$.   In any case, we recognise that in photosynthesis the reaction takes place in several stages, and that the first stage consists in a reduction process, which is made evident by the evolution of oxygen.

Water-plants such as Elodea serve also for investigations into the conditions under which photosynthesis takes place.

189. Select a healthy shoot of Elodea, make a clean cut across its stem, wipe the cut end gently with blotting-paper, and brush a layer of collodion over it.   When the collodion has set, make a small hole in the film by means of a needle, tie the shoot, cut end upward, to a glass rod, and submerge it in a large beaker of water. Using the rate at which bubbles escape from the cut end as an index of photosynthetic activity, determine the influence of light-intensity on the rate at which this process goes on.   This may be done by counting and comparing the rates at which bubbles are given off when the apparatus is placed at varying distances from a well-lit window.   A better method consists in the use of an incandescent light in a dark room and the determination of the rates of the evolution of bubbles when the apparatus is placed at known distances, *e.g.* 1 ft., 2 ft., and 4 ft., from the source of light. We are thus able to demonstrate that photosynthesis becomes very slow when the plant is feebly illuminated, and that it increases as the plant is brought into brighter light.

A similar apparatus may be used to demonstrate the results of Exp. 175, viz. that carbon dioxide is indispensable for photosynthesis.

190. For this purpose fill a large beaker with water which has been so thoroughly boiled (see Appendix B) as to expel the carbon dioxide which it held in solution. Having submerged in the beaker the glass rod carrying the Elodea shoot, pour on to the surface of the water a layer of oil in order to check absorption of carbon dioxide from the air. Using the apparatus of Exp. 189 as a control, set the two beakers in a good light, and determine, by bubble-counting, that the rate of evolution of oxygen is considerable in the case of the plant in unboiled, and practically nil in that of the plant in the boiled water.

If the apparatus of Exp. 188 is placed in a dark room, it will be observed that no gas accumulates in the test tube, and thus our conclusion is confirmed that photosynthesis goes on only in the light.

Lastly, we have to enquire into the part played by light in photosynthesis.

Every photographer knows that certain chemical changes are set up or accelerated greatly by light. Hence it is not altogether surprising that the chemical process involved in the synthesis of carbohydrates is set going by light. Light is a form of energy (radiant energy), and hence is capable of doing work. The fact that leaves are green indicates that they possess the power of absorbing certain rays of light. The " white " light which falls on the leaf consists of an assemblage of ether-waves of different lengths. Since those of greater wave-length, when separated from the others, for example, by a piece of red glass (which absorbs all but the red rays) produce in us the sensation of *redness,* we speak of them as " red " rays. Rays of shorter wave-length, when separated sufficiently from the others, *e.g.* by a piece of blue glass, which absorbs most of the latter, give rise in us to a sensation of *blueness,* and so may be called "blue" rays.   By means of a prism of glass, the ether-waves which make up white light can be separated into groups which form a series of increasing wave-lengths from red to orange, yellow, green, blue, indigo and violet; beyond the violet are ultra-violet rays, to which our eyes are blind; beyond the red are infra-red rays, which are likewise without optical effect (see Bibliography, 14, 15).

We require to determine which parts of the visible spectrum, as this series of colour-producing rays is called, are concerned in photosynthesis.

If a spectroscope is available, we first ascertain which parts of the spectrum are absorbed by an alcoholic solution of chlorophyll made by putting chopped grass or other leaves in a stoppered bottle, adding methylated spirits and keeping the bottle in a dark place till the green colour is extracted.

191. Pour the solution into a test tube—or better, into a narrow glass vessel with parallel sides—and bringing the vessel into a good light, *e.g.* in the open, held up against a white cloud, we look at it through the spectroscope.   In the spectrum, as seen through the green liquid, dark bands appear.   These bands are called *absorption bands,* since they are due to the absorption of certain rays by the green chlorophyll solution.   The bands are in two groups, one set in the red end, the other in the blue-violet end.   Hence we conclude that chlorophyll is green not because it absorbs all but the rays which give rise to the sensation of greenness, but because, though it lets pass also some of the rays at the red and blue-violet ends of the spectrum, the amounts of the red and blue rays which it transmits are not enough, swamped as they are by the large quantity of green rays to which chlorophyll is more transparent, to produce an optical effect on our eyes. The eye and brain make a rough optical judgment, and tell us the partial truth that leaves are green.

Now, if the energy of light is used by the chlorophyll, we shall expect to find that those rays which are absorbed by the chloroplasts are the rays which supply the energy for photosynthesis. In order to demonstrate this, we use one or other of the following methods :—

192. Double bell-jars (Appendix B) are filled, one with an ammoniacal solution of copper sulphate (Appendix A), the other with a solution of potassium bichromate (Appendix A) ; each solution being of such strength that the former blocks out the red end of the spectrum and the latter the blue end. The bell-jars are placed over two similar plants, clover, bean, etc., which may have been kept in darkness for a day or two until samples of the leaves show, on testing, no starch reaction. The plants covered by the bell-jars should be in a good light (not in direct sunshine), and soil should be piled about the bases to prevent light from reaching the plants except through the coloured liquids. After two days, test leaves of both plants for starch, when it is found that those from the plant subjected to blue light contain no starch, and those from the plant grown in red light contain considerable quantities.

193. A better method of conducting this experiment involves the use of "dark boxes," made each large enough to contain a small plant and provided with a light-proof removable top and also with a large opening in one side, into which a " window " or coloured screen may be fitted (Fig. 12) and Appendix B. The screens are made as follows :—

Red. Good red photograph glass.

Green. Signal-green glass.

Blue. Blue stained glass, a piece of transparent plate-glass, and enclosed between them two or more blue gelatine sheets (Appendix B).

The quality of the light transmitted by the screens having been determined spectroscopically, they are fitted each into the opening of a dark box. Pot-plants, *e.g.* clover, Tropæolum, tomato, etc., are placed, one in each box. The boxes are exposed to a north light (to avoid direct sunlight) and leaves of the plants which they contain are tested at daily intervals for starch.

As the results show, much starch accumulates in the leaves of the plant exposed to red light, little or none in those in the plant in blue light, and none in the leaves of the plant in the green light. Hence we draw the conclusion, verified by more accurate methods, that the light—particularly that of the red end of the spectrum—which is absorbed by the chlorophyll, is that which supplies the energy for photosynthesis.

The essential conclusion to be drawn from our work on nutrition is, that the cells of the adult plant, like those of the seedling and of animals, obtain their carbo-hydrate food-materials in the form of sugar; but that, whereas the seedling and the animal obtain their supply of carbohydrate food-material ready-made, the independent green plant manufactures it from the inorganic raw materials, carbon dioxide and water; that the chloroplasts constitute the apparatus which carries on this photo-synthesis; and that the energy for it is supplied by the sun.

Hence, when George Stephenson said "it is the sun that drives the steam-engine," he stated the bare truth. For the coal used in the furnace consists mainly of carbon, the remains of carbohydrate material formed by bygone races of plants. By the activity of their chloroplasts, and by the radiant energy of the sun, ancient plants con-structed this carbohydrate. Much of that which they formed was used in carrying on the work of their lives. To serve this purpose, it was oxidised in the respiratory pro-cesses to carbon dioxide and water, that is, into the original raw materials whence the carbohydrate was derived. Much of it also was used in plant-construction, incorporated with protoplasm and excreted as cellulose, etc. These substances, remaining in the dead plant, underwent slow changes of decay, lost oxygen and hydrogen, and the de-oxidised carbon-compounds were left finally as the chief constituents of seams of coal.

Just as animals and just as seedlings derive directly or indirectly their carbohydrate from the activity of in-dependent green plants, so every non-green cell of a green plant needs must rely on the green cells for its supplies. So it follows that the large number of non-green plants, the mushrooms and toadstools, the bacteria—with one or two exceptions—the parasitic non-chlorophyllous plants like the dodder (Cuscuta europæa, etc.), the saprophytic plants like the bird's-nest

orchis (Neottia nidus-avis), which live on decaying vegetable matter, all absorb their carbon in the form of
organic compounds derived from the partial decay of
the direct or indirect remains of green plants.   In short,
all life is a monument constructed of materials which the
green plant alone is able to manufacture.

With respect to the synthesis of organic nitrogen compounds, which, as we know, are integral parts of the
living plant and animals, our knowledge is but fragmentary.   How the plant manufactures, from the nitrates and
other salts, the organic nitrogen-compounds which serve
as food-materials for the construction of proteins, remains
for the botanists and chemists of the future to discover.

CHAPTER XII.

The majority of the experiments which we have performed hitherto have had for their purpose the discovery of the processes whereby the living plant obtains its food and utilises that food both constructively and destructively. We have learned that the green plant possesses the power of building up its food-substances from inorganic raw materials, and that, having done so, it utilises them in the same way as an animal utilises the similar food-substances which it obtains ready-made. But, in the course of these studies, we have had frequent illustrations of the fact that the plant possesses the power of putting itself in the position most appropriate for the carrying out of its work. We have seen that its behaviour is not haphazard, but purposeful. It has work of different kinds to do, and it not only so disposes its members that they may carry on their several tasks to the best advantage, but it possesses also the power of responding to changes in external conditions in a no less remarkable manner. Under the influence of changed external conditions, the course of development of the organs may undergo such modifications as fit them the better to carry on their work in the new conditions. As an illustration of this fact, we may refer to the differences in the mode of development of the root-hairs of plants grown in water and in moist air or earth (p. 94). Many other illustrations might be given of this power possessed by a plant of responding to external conditions by modification of its habitual structure. Thus the reduction of leaf-surface so characteristic of xerophytic plants (p. 148) is evidently of the nature of an adjustment, serving to reduce the risk of damage from drought. It is true that, in many xerophytes, the reduced form of leaf is a racial habit, and, in such cases, even though the plants are grown in a moist atmosphere, the habit of producing reduced leaves, or no leaves at all, is too strongly engrained to be abandoned now. But, in some cases, it is not difficult to prove that the structural habit is not so fixed, and that change of external conditions may suffice to modify it.

194. For this purpose raise seedlings of gorse (Ulex europæus) in pots. As soon as the seedlings appear, place one pot under a bell-jar standing on sphagnum moss, the edges of which, projecting beyond the rim of the pot, are kept wet. Place another pot in a green-house or prick the seedlings out in the open ground. Observe, by comparing seedlings with adult branches of gorse, that whereas the former possess well-developed foliage leaves, the leaves of the adult shoots consist of small scales. Note also that the green assimilatory tissue occurs, in the adult plant, in the thorny shoot-system. Dry seedlings and branches of the full-grown plant, and add them to the museum. Compare at fortnightly or monthly intervals the plants grown in the moist atmosphere of the bell-jar with those grown under ordinary conditions, and observe that the former retain their typical leaves for a much longer period than the latter plants. Such a mode of response—by structural modification—to changed environmental conditions may be called a *morphogenetic* response.

195. Further confirmation may be obtained by growing xerophytes, which show adaptive characters, such as a "rosette" habit, under moist conditions for some months, *e.g.* under bell-glasses in a fernery. The result of an experiment with a Sempervivum is shown in Fig. 26.

The plant (B), grown under a bell-glass in a damp green-house for twelve months, shows marked differences of habit and structure from the originally similar plant (A),

which grew under normal conditions. The stem of B
has developed well-marked internodes, the leaves are less

A

B

Fig. 26.

A, plant of *Sempervivum sp.* under normal conditions; B, similar plant grown for
12 months under a bell glass, in moist air in a green-house.

succulent, and the waxy bloom, which covers the normal
leaves, is almost absent.

We thus learn that purposeful adjustments may be effected in different ways. Either the adjustment may be left to the enterprise of the individual plant or it may have become so deeply engrained as a racial habit that the individual plant is powerless to alter the habit of the race, even when external conditions urge it, as it were, to make the effort.

Again, the leaves of a tree, such as the beech, seem, except for details of size and shape, to be very much alike. If, however, the tree is growing on the edge of a wood, we may satisfy ourselves that the leaves on the shady side present marked and constant differences in structure from those on the sunny side.

196.* Cut cross-sections of the two types of leaf, and observe that the palisade parenchyma is more developed in the sun-leaf than in the shade-leaf. Accepting the view that the palisade parenchyma plays the part of a screen, protecting the more actively assimilatory spongy parenchyma from excessive sunlight, we learn that the plant possesses the power of automatically modifying the development of its individual leaves according to the situation which they occupy. It requires no argument to convince us that these adaptations on the part of the members of a plant are not mechanical results, dependent merely on the physical properties of the plant's tissues, but physiological results dependent on the vital activity of the plant. Kill it, and all such adjustments are impossible.

The property of the living substance (protoplasm) which enables it to respond purposefully to change of environment, either by modifying the development or by changing the positions of its members, is called *irritability*.

We devote ourselves in this concluding chapter to a brief study of the irritability of plants.

The subject is so large that to treat it properly would require not a chapter, but a book. We shall, therefore, not attempt to deal with it in full, but rather limit ourselves to a careful study of a few examples of the manifestation of irritability in plants.

In the case of the higher animals, we know that responses to external changes are carried out by the agency of a specialised nervous system. If our shoe pinches, we take it off. The removal of the shoe is the consequence of a complex series of nervous events. The pressure of the shoe sets up changes in special organs in the skin of the foot. These special organs—tactile organs —contain the endings of nerves. The tactile organs are perceptive organs—their function is to "perceive" changes of pressure. The agent which calls forth the perception— in the case under consideration, the pressure produced by the tight shoe—we call a stimulus. As a consequence of the stimulus, changes are set up in the organs of percep- tion. We say they "perceive" the stimulus. Thus excited, the perceptive organs induce changes in the nerve- endings which they contain. As a consequence of these changes, others, which we call nervous impulses, are propagated along the nerve to the central nervous system. These nervous impulses produce definite effects in the nervous tissue of the central nervous system. Thus they may set up in the brain changes which affect our con- sciousness and give rise to the sensation of pain. They set up changes also in other parts of the central nervous system. As a consequence of these changes, nervous impulses travel along special nerves to the muscles of foot, and we wriggle our toes as though to withdraw them from the tight shoe. Finding this unavailing, and not liking the sensation of pain, we relieve it by another complex set of nervous and muscular acts, which result in the removal of the shoe. This sounds very complicated, and so it is. What we have to do is to endeavour to unravel such a nervous tangle as this, and to discover if possible of what simple nervous constituents it is composed. We may remark, at once, that, though of the highest importance to us, the conscious part of the nervous series of events may be dismissed as an extra, and not fundamental, part of the nervous complex. If we went to bed with our shoes on, and they began to pinch after we had fallen asleep, our toes might wriggle just as, or even more purposefully than, if we were awake; so that in the morning we might find that we had kicked our shoes off in our sleep. We are, therefore, prepared to find that among the lower animals—in which consciousness is not apparent, or at best appears in but fitful gleams—changes in environment, that is to say, stimuli such as pressure, may result in purposeful operations without the co-operation of consciousness. Such purposeful operations, carried out by the nervous system, we may call *reflex* acts. Now, as we go down in the scale of life, we find animals in which the nervous system—brain, spinal cord, and nerves—becomes more and more rudimentary. Yet even the lowest animals, in which no recognisable nervous system exists, are very well able to take care of themselves. They respond, like Euglena and Chlamydomonas (p. 84), to the light, they seek their food, ingest it, avoid objectionable objects, and take up definite positions in the world. Therefore, we conclude that reflex acts may be carried out without any elaborate apparatus of brain and nerve; that, as irritability is a fundamental property of protoplasm, as weight is of matter, so the purposeful reactions of organisms depend on the living protoplasm ; and that, therefore, every living organism, whether plant or animal, is capable of reflex acts, that is, is capable of responding to stimuli by movements or other operations of adjustment. Our work, therefore, comes to be the experimental examination of the reflex actions of plants.

We commence this investigation by finding out what are the chief stimuli to which plants respond; or, in other words, what are the external, environmental changes with respect to which plants show, most generally, powers of adjustment? In its pilgrimage through life the plant requires sign-posts, lest it should lose its way. What are the "sign-posts," how does the plant "read" them, and how is it able to interpret them in the interests of its life-journey? The facts which we have learned in previous chapters enable us to give at once a partial answer to the first question. The first act of the radicle on emerging from the seed is to turn vertically downwards, that is, to put itself right with respect to the line in which gravity acts. If we look at ivy growing on a wall, we shall observe that practically every leaf exposes its upper surface to the light. Closer inspection shows us that this entails in many cases an elongating and a bending of the petioles. This uniform disposition of the leaves cannot be due to accident; we may be sure that it is the result of a response on the part of the leaves to the stimulus of light. *Light* and *gravity* serve, therefore, as important stimuli in response to which the members of plants take up their several proper positions. In order to investigate the modes of response of the plant to the stimulus of gravity, we proceed as follows :—

197. Grow seedling beans, sunflowers, etc., in glass-fronted observation-boxes. When the roots are an inch or so long, trace their outlines on a piece of thin tracing-paper fixed to the glass-front. Turn the box so that it rests on one end. After two or three days, note that the root-tips are once again pointing downward. Trace the outlines of the roots, observe that they have grown and that the line of growth, instead of being, as before, vertically downward, is curved so that the upper side of the root is convex. It follows from this observation that the adjustment of the root, effected by its response to the stimulus of gravity, is brought about by a curvature of a certain part of the root. For this reason we may speak of the response as a *tropism* (bending) or *tropistic* movement, and since the stimulus which evoked the response was that of gravity, we describe the movement as *gravitropic,* or, to adopt the more usual term, *geotropic.* Now, compare the region of curvature of the roots with the records of the region of elongation (Exp. 71). Observe that the geotropic curvature is most marked in the region where growth in length is greatest, and that the region of curvature coincides closely with that of elongation. If the results are not convincing, grow young seedlings in an observation-box. Select several with short, straight roots, mark them in the manner described in Exp. 71, and replant them.

198. Then repeat the experiment of turning the box on its side and determine that the region of curvature coincides with that of elongation.

We thus learn that the geotropic response of the root is effected by means of a growth curvature. That is, in consequence of the changed position of the root with respect to the vertical line along which gravity acts, growth, which was previously as rapid on one side of the root as on the other, has become more rapid on the upper side of the region of elongation than on the under side.

Observe next that, as a consequence of the movement, the root-tip is brought to point once again vertically downward. When this is effected, the curvature of the growing region ceases and the root continues to grow in a straight line. The gravitational stimulus acts along a vertical line, and the root, in response to the stimulus, places itself in a definite position with respect to that line. The stimulus may, therefore, be described as a *directive stimulus,* and the reaction also as a *directive reaction.* In this case, the root grows in the direction (towards the centre of the earth) whence the stimulus proceeds. We, therefore, describe the response as one of *positive*, or + *geotropism*, and say that the main root of a plant responds by a positive geotropic movement or curvature to the directive stimulus of gravity. We have a ready means of demonstrating in a striking manner that the geotropic curvature of the root is a growth curvature; for we know that after a given region of the root has ceased to grow in length, the further elongation of the root is effected by a region between it and the root-tip (p. 88). Therefore, we turn the observation-box back once again into its normal position, and observe that, after some days, when the root-tips have been swung once again to point vertically downward, the old curvature, induced when the box was first turned on its side, is still visible. That part of the root which was in course of elongation when the box was originally turned on its side had in the meantime ceased to grow, and hence it failed to be affected by the new stimulus produced by changing again the position of the root with respect to the line of action of gravity.

During the course of this experiment, it will not have escaped observation that the shoot also responds by directive movements to the stimulus of gravity. When displaced from the vertical, the shoot curves geotropically; but, unlike the root, it curves in the opposite sense, that is, so that its apical region is directed away from the centre of the earth. In other words, the shoot-axis is *negatively* geotropic. No better illustration than this differential behaviour of the root and shoot of one and the same plant

could be found to convince us that, though gravity gives
the signal, the plant determines, with reference to utility,
the mode of response of its several members. Such
responses are therefore purposeful; but by virtue of what
powers the plant is capable of this discrimination no one
knows.

When seedling plants in the observation-boxes have produced lateral roots, we investigate by the simple device
of turning the box on its side this wonderful power of
differential response to one and the same stimulus.

199. Trace the outlines of the lateral roots and observe that they stand out from the main root at wide
angles—more or less horizontally. On turning the box
on its side, we discover, after some days, that the lateral
roots, like the main root, exhibit geotropic curvatures, and
that, as a result of these curvatures, the tips of the lateral
roots stand out once again more or less horizontally, and
make the same angles with the main root as they did at
the beginning of the experiment.

Interpreting these results in our technical terms, we
say that lateral roots are *transversely geotropic.*
Though their structure is, so far as the eye, aided by
the microscope, may judge, identical with that of the
main root, yet they respond by a different directive
movement to the stimulus of gravity. As soon as the
tip of the main root is brought into the plumb-line—
pointing vertically downward—the root ceases to respond
to the stimulus of gravity. As soon as the tip of the
lateral root makes its proper, wide angle with the main
root, the tropistic curvature ceases, and the lateral root
grows on in a straight line. The position of equilibrium,
or rest-position, with respect to gravitational stimulus is for
the main root, the vertical, for a lateral root, a horizontal
position. Once in their respective rest-positions, gravitational stimulation induces no curvature in these organs.
But an extremely small displacement from the rest-position
suffices to set up excitation in the root, and to induce a
geotropic response whereby it is brought once again into
its rest-position.

Yet another mode of behaviour is exhibited by the
tertiary roots. In seedlings which have grown in a deep

observation-box long enough to develope a complete root-system of tertiary, as well as primary and secondary roots, the former are seen to run in any direction, some upwards, some downwards, and some more or less horizontally. In other words, the tertiary roots fail altogether to respond by directive movements to the stimulus of gravity, that is, they are *ageotropic* (non-geotropic). They rest in any position undisturbed by gravitational stimulus.

If we look at the root-system of a well-grown seedling, *e.g.* through the glass side of our observation-box, we see that its members explore in a thorough and systematic manner the soil in which they lie. The lateral roots each occupy a section of the mass of soil through which the main root runs, and the branches borne on the lateral roots pass in all directions through these layers of soil. Thus the root-hairs of the several roots are brought into contact with different regions of the soil, and so, instead of competing with one another for supplies of water and mineral salts, they each have their special spheres of influence. It is evident that this happy state of affairs is due, in large measure, to the different modes of response by primary, secondary, and tertiary roots to the stimulus of gravity. These responses are, therefore, not haphazard and meaningless, but purposeful : that is, they make for the efficient performance of the functions of the root-system. That the ordered and purposeful behaviour of the members of the root-system with respect to the stimulus of gravity is controlled by the root-system as a whole, we prove in the following manner :—

200. After tracing the outlines of its root-system, re-move from the observation-box a bean with a well-developed root-system. Cut away an inch or more of the main root, and replace the plant in the box, embedding its root-system in moist but not wet sphagnum moss. Observe, after some days, that one of the lateral roots has curved so as to bring its root-tip to point vertically downward. As when a captain falls, the next in rank assumes command and undertakes the duties of leadership, so here, the main root destroyed, a lateral takes its place and behaves like its pre-decessor in command. Note, further, that the tertiary roots, borne on the lateral root now promoted to take the place of the main root, become capable of geotropic response, and come to lie more or less horizontally. They also have received a step in rank, and pass from the ageotropic to the transverse geotropic state.

201. Similar behaviour may be seen or induced in the shoot of such a plant as the Scotch pine (Pinus sylvestris). In this plant, the terminal bud builds up year after year the upright main shoot, and the lateral buds, borne on the main shoot just below the terminal bud, give rise to lateral branches. By removing the terminal bud, however, it may be shown that one of the lateral buds, instead of growing out more or less horizontally to form a lateral shoot, developes into an upright growing branch, and thus replaces the lost leader. Though the plant consists of a number of different members, each with its proper position and special behaviour, yet the ordering of position and behaviour are determined, not by the members independently, but by the organism as a whole. It is by virtue of this that the plant behaves as an individual and not as a collection of independent members. We may say that, by reason of this control, the plant governs its members, whereas, without it, anarchy would reign among them. Geotropic responses may be exhibited not only by members of the root-system and by the shoot, but also by other members of the plant, *e.g.* the flower-bearing axes. Thus if flower buds of Narcissus are observed, the flower-stalks are seen, each to be bent sharply just below the ovary, so that the corolla tube stands out horizontally.

202. To demonstrate that this position is the consequence of a transversely geotropic response, cut off a flower-stalk of a narcissus or daffodil, fix it in water in a test tube, and incline the test tube (supported by means of a retort stand) so that the flower is directed downward. Fix a sheet of lightly-ground glass vertically on one side of the apparatus and make a sketch of flower-stalk and flower in order to have a record of their positions. At intervals, retrace the outlines and determine that, in the course of twenty-four hours, the bend in the peduncle (flower-stalk) has decreased so much that the flower stands out once again horizontally.

It is noteworthy that, in this case, the region of bending

is very localised. Similar localisation of the region capable of movement in response to gravity is to be met with in other parts of plants, for example, in the haulms, or jointed stems of grasses.

203. Observe in the growing haulms of any tall grass the swellings (pulvini) at the lower nodes. Cut a haulm, make a series of india-ink marks on either side of a pulvinus, and measure the distances between the marks. Fix its cut end in a box, *e.g.* a biscuit box, containing moist sand. Place the box so that the haulm is horizontal, and trace by drawings on a sheet of ground-glass the progress of the shoot to a vertical position. Note that this is effected by a sharp curvature of the pulvinus. Remeasure the distances between the india-ink marks, and observe that they are now greater on the lower side and smaller on the upper side of the pulvinus than when the marks were made. We thus discover that the negative geotropic response has been effected by the growth in length of the lower side of the pulvinus, and that, accompanying this increase, there has been a decrease in length on the upper side of the pulvinus.

The final proof that the curvatures of the kind we have been considering are responses to the stimulus of gravity may be obtained by so rotating a plant, that first one side, and then another is directed downward. This is effected by fixing it to a horizontal axis, which is kept revolving at a slow rate by clock-work or other means. The cheapest form of this apparatus—which is known as a *klinostat* (Fig. 27)—consists of a clock movement, with which is connected a horizontal axis bearing at its end a holder in which a block of peat or a light pot may be clamped. Soaked seeds of mustard, bean, etc., are fixed to the moistened block of peat and the whole apparatus stood under a dark box or beneath a bell-jar covered with black cloth. The clock-work is set going and as the seeds germinate, the seedlings are revolved about the horizontal axis. A root, emerging from the seed in any direction, finds itself in a novel condition with respect to gravity. Each side in turn is exposed to gravitational stimulus, and each such stimulus tends to induce a geotropic curvature in a definite direction. Therefore, there

is no more reason why a root should bend in one direction
rather than in another, and so the roots grow in any
direction. At one time, one side of the root faces downward and its opposite side upward. The root, therefore,
tends to bend positively geotropically; but, before the
movement can be carried out, the side which was below
is above, and hence it tends to respond in exactly the
opposite sense. Subject to such contradictory gravitational

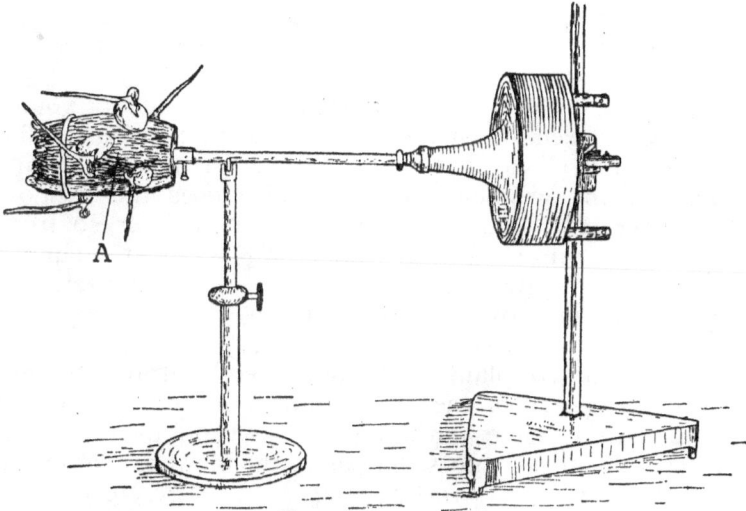

FIG. 27.—SEEDLINGS OF BROAD BEAN (VICIA FABA) GERMINATING ON A
KLINOSTAT.

A, a block of moist peat or moss on which the seeds are planted.

stimuli, the movements of response cancel out. Like a
person beset with a rapid series of contradictory orders,
the root, in setting out to execute them all, achieves none
(cf. Fig. 27). The shoot shows also a like behaviour, and,
when attached to the klinostat in the dark, grows in any
direction indifferently.

We turn now to investigate the tropistic responses
which light calls forth in the plant. If we observe a plant
growing in a corner of a room or in a window, we may

note that the positions taken up by its axis and leaves are different from those occupied by these members in a plant growing freely in the open. Again, if we look at a group of daffodils or narcissus, we may observe that the bent heads of the flowers are all directed towards the sunny side. Thus we are led to suspect that light, like gravity, plays a part in determining, not only the normal positions assumed by the axis and leaves, but also in calling forth, under certain circumstances, directive movements, which result in adjustments of axis and leaf to special light conditions. Or, to use our technical terms, light may not only induce, as we have learned, morphogenetic, but also tropistic responses.

204. That this is the case we prove by putting young plants, *e.g.* seedling sunflowers (pot-plants) in boxes like those used in Exp. 192. One box is made quite light-proof; in the wooden shutter which closes the window of another, a circular hole is bored an inch or so from its top edge; in a third, a similar hole is cut at a similar distance from the bottom edge of the shutter. A fourth plant is grown in the open, in order to serve for purposes of comparison. Whilst the experiment is in progress, the control plant is photographed or drawn so that we have a record of the disposition of its axis and of its leaves. After some days, an examination of the plant grown in darkness shows that, as in the plant grown in the open, the stem points vertically upward. Therefore, apart from any response which the axis makes to light-stimulation, it responds by negative geotropism to the directive influence of gravity. On removing the windows from the remaining two boxes, we discover that the plants contained therein exhibit remarkable tropistic curvatures, the effects of which have been to cause the tip of the stem to point in each case toward the opening through which the light entered the box. Thus we learn from the experiment that the stem of a plant such as the sunflower, when illuminated from one side, responds to unilateral light by a *positively phototropic,* or as it is sometimes called, a *positively heliotropic,* movement. In the case of the plant illuminated unilaterally and from above, a comparatively small curvature has sufficed to bring the young part of the axis into a line parallel with the direction of the light; but, in the case of the plant illuminated unilaterally from below, the curvature necessary to bring the axis parallel with the direction of the unilateral light has been very considerable. Once it has been brought so as to lie parallel with the direction of the light, and pointing towards the light, no further curvature takes place.

The phototropic curvatures exhibited by these plants are interesting also from another point of view. In the performance of the phototropic responses, the young part of the axis has been brought out of the vertical, and hence has been subjected to the stimulus of gravity. Now, we know from the behaviour of the plant grown in darkness that the shoot is negatively geotropic. Knowing this, we might have expected that the unilaterally-illuminated plants would have struck a compromise with respect to response to light, and response to gravitational stimuli, and have taken up a half-way position. We learn, however, from the behaviour of the two unilaterally-illuminated plants that this is not the case. The axis, though as soon as it has curved away from the vertical it is stimulated by gravity, ignores the gravitational stimulus completely, and pays heed only to the unilateral light-stimulus. If, instead of light and gravity, we act on the plant by means of two mechanical forces, e.g. by attaching strings to the axis and pulling one vertically upward and the other obliquely downward, the extent of the curvature and, hence, the final position depends on the relative pulls we exert; and supposing the axis to be quite freely bendable, we could calculate beforehand the position it would occupy as the resultant of its reaction to known mechanical forces. But, with the case of stimuli and responses by the living organism, the resultant position cannot be thus calculated, nor need it be intermediate between the directions in which the two stimuli act. The only way by which we might be able to predict the physiological resultant of two simultaneous directive stimuli, would be by considering which stimulus it would be most useful for the plant to obey. In the case under consideration, it is evident that negative geotropism is a useful guide to the axis, enabling it, when buried underground, to reach the light. But it is also evident that, once it has emerged from the ground, light is the better guide. The functions of the leaves can only be carried on properly in a good light, and hence, when the light is none too good and comes from one side only, the plant, to do its work most efficiently, must put itself into the best position. To obtain the most favoured light treatment, it must resign itself exclusively to the guidance of light. This it does in the case we have just considered.

Examination of the plants grown in unilateral light shows us that the changes of position which this treatment has induced are not confined to the axes, but extend also to the leaves. In the control plant grown in the open under uniform illumination, the leaves lie more or less horizontally face upward to the light. But, in the boxes illuminated from one side, the leaves have turned themselves in such a way that they no longer lie in approximately horizontal planes. When we look at them through the holes—removing the tops of the boxes to admit light for their inspection—we see that, by twisting of their petioles, the leaves have so placed themselves that they present their faces (upper sides) to us; that is, the leaves lie athwart the line of light. We conclude, therefore, that the leaves of plants are *transversely phototropic*. By the use of the klinostat, we are able to demonstrate this conclusively.

205. Dig up a plant of Ranunculus ficaria (Lesser celandine) and enclose its roots in damp moss. Note that its leaves curve downward. This curvature, it should be observed, is not the result of response to gravitational or light stimuli, but is due to stimuli which come from the plant itself (cf. p. 223). Fix the plant on the klinostat with its axis horizontal and parallel to the spindle of the apparatus, and place the klinostat so that its axis points directly to the window of a room. After some days, look at the plant from the window side, and note that its leaves have, as the result of transverse phototropic response, placed themselves so that the faces of their blades are at right angles to the direction of the light.

By repeating the experiment of growing plants behind coloured screens, we obtain evidence that the

rays of light which serve as the stimulus to photo-
tropism are those of the blue-violet end of the spectrum.
Grown in red light, the plant shows no phototropic curvature, whereas, when grown in blue-violet
light, it responds as markedly as in white light. Thus,
for different purposes, the green plant makes use of
the light from the two ends of the spectrum. The red-
orange rays are used chiefly in the process of photosynthesis, but not to the complete exclusion of those from
other parts of the spectrum. The blue-violet rays serve as
beacons, in response to which the plant executes its purposeful, phototropistic, reflex movements.

It is easy to show that, though gravity and unilateral
light play extraordinarily important parts in guiding
the members of the plant to their proper positions,
other agents may also induce tropistic curvatures.
Thus it happens not infrequently that plants, *e.g.*
cucumbers, etc., grown in a very moist atmosphere,
send up some of their roots from the soil into the air as
though the roots were attracted by the moisture. Again,
it happens fairly often that, when a stoppage occurs
in field drain-pipes, they are found to be blocked with the
abundant roots of some plant growing in the neighbourhood. That the roots of plants actually possess this water-
finding power, we may illustrate by the following experiment :—

206. Place several layers of moist moss in a small sieve,
plant seeds of radish or mustard, etc., in the moss, and
hang the sieve obliquely over a wide-mouthed vessel, at
the bottom of which calcium chloride is placed in order
to dry the air. Take care that the moss is kept moist.
Observe that the roots of the germinating seedlings,
instead of growing vertically downwards, curve toward
the moisture contained in the moss, and may indeed re-enter
the sieve and thus escape from the dry air.

We learn from the foregoing series of experiments
that it is not by a sequence of happy accidents that
a green plant disposes its roots in the most suitable positions in the soil, or raises its stem vertically and spreads
its leaves to catch all the sunbeams possible. These
movements are for the most part the result of tropistic

responses to stimuli. Among the directive stimuli, those of gravity and light are the chief. The stimuli are perceived by the plant, and the mode of its reaction is not inevitable, but purposeful. This member reacts in this way, and that, in that way, and each reaction serves the end of disposing the member concerned in that position in which it may best carry on its work.

We must now endeavour to resolve those tropistic reflex movements into their component parts. If we refer to the drawings made to illustrate the mode of geotropic curvature of the root, we find that curvature occurs in the region of elongation of the root, and that it continues till the tip of the root is carried into the vertical line. Then it ceases. Evidently, therefore, the root-tip, though it does not bend itself, counts for something in a geotropic response. In order to prove that this is the case, we repeat a famous and simple experiment made long ago by Charles and Francis Darwin.

207. Germinate a dozen beans, and, when the roots are about 1-1½ inches long, remove, by as transverse a cut as possible, the root-tips from six of the seedlings : the portions removed should be each about one millimetre in length. Lay the amputated and unamputated seedlings, with their roots horizontal, in a germinator. After a day, examine them, and observe that, whereas all the intact roots have begun to curve geotropically so that their tips point downward, the decapitated roots have not curved geotropically, though some, particularly if the cuts were not made strictly transversal, may show irregular curvatures.

The behaviour of the decapitated roots suggests the question, why does the removal of the tip of the root prevent geotropic response? Measurement of the decapitated roots on successive days proves that they grow in length and that therefore the operation has not sufficed to kill them. Why then have they not curved? Why is it that the decapitated root fails to perform the geotropic reflex? The failure might be due to one or other, or both, of two causes. Either the root loses, with the removal of its tip, the power of perceiving the stimulus of gravity, or it loses the power of responding to the stimulus or both these powers are lost.

In order to be in a position to deal with this question experimentally, we will first investigate a little more closely the mode of behaviour of a root when it is subjected to gravitational stimulus. As we must have observed already, a root does not respond by immediate curvature to the stimulus of gravity. Before the stimulus can produce its effect, it has to act on the root for a certain length of time.

208. This period, called *presentation time,* may be determined roughly by exposing roots to the stimulus of gravity, *e.g.* by placing them horizontally for different lengths of time and then putting them on the klinostat. Since, as we know, the root of a plant revolving about a horizontal axis shows no definite curvature response to gravitational stimulation, we infer, if the roots revolving on the klinostat show well-marked curvatures, that these curvatures are the belated responses to the gravitational stimuli to which they were subjected before they were put on the klinostat. As a result of the experiment, it is found that roots which were exposed to the stimulus of gravity for a very brief time exhibit no definite curvature after they are put on the klinostat, but those which were exposed in a horizontal position for a longer time execute definite curvatures.

As estimated by this method it has been found that the presentation time—that is, the least time during which a root must be exposed to gravitational stimulation for the stimulus to induce a tropistic response—is, in the case of certain plants, about twenty minutes.

Again, it will be evident from previous experiments that a root does not react by curvature till some considerable time after it has been subjected to gravitational stimulation. Just as a certain interval of time (presentation time) elapses before the stimulus impresses itself effectively on the root, so a yet longer interval elapses before the results of stimulation are made manifest in the geotropic movement. The time which elapses between stimulation and response is called *reaction time,* and may be estimated by ascertaining the length of time which elapses between exposure to an efficient gravitational stimulus and the commencement of the geotropic response.

209. In order to make a rough determination of reaction time, lay a series of roots horizontally, *e.g.* pinned each to a cork in a glass vessel placed on its side. Having fixed the cork, and also the bottle, place a sheet of glass in front of the bottle and trace thereon the outline of the root. This done, take a small piece of wood, *e.g.* the lid of a cigar box, make a slit about two inches long and half an inch wide and fix by shellac a thin taut thread, or fine wire, so that it passes across the middle of the slit parallel to its longer sides. Fix the board in a clamp so that the cross-wire is just between the root and the drawing. By observing at intervals, determine how long a time elapses before a distinct geotropic downward curvature of the root occurs. Having ascertained that the reaction time for a bean root is about an hour and a half—our method is only a very imperfect one—we apply our knowledge to the solution of the problem of the meaning of the behaviour of the decapitated roots of Exp. 207. The problem is to ascertain experimentally whether a decapitated root is or is not capable of geotropic curvature.

210. To determine this point, lay six bean seedlings horizontally, and leave them thus exposed to the stimulus of gravity for a time less than their reaction time, *e.g.* for about an hour and a quarter. After noting that no geotropic curvature has occurred, remove the root-tips, as in Exp. 207, and place the seedlings with their roots vertically downwards in a germinator. After they have been in this position for half a day, take them out one by one and observe that the effect of their previous sojourn in the horizontal position is manifested in each by a curvature, the side which was uppermost when the root was in the horizontal position being now convexly bent. A decapitated root, therefore, is capable of carrying out a geotropic response in obedience to a stimulus applied to it when it was intact. Consequently, we infer that the failure of the decapitated root to respond by curvature to gravitational stimulation is due, not to a failure of the motor apparatus, but to a failure to *perceive* the stimulus. Whence it follows that we must regard the root-tip as an organ for the perception of the stimulus of gravity. This conception of a definite localisation of the power of perception of a stimulus, though it may appear novel with respect to plants, is a commonplace with respect to animals. Most members of the animal kingdom possess specialised organs of perception, eyes for the perception of the stimulus of light, and so on. Therefore, we shall not be surprised to find a similar—if less precise —localisation of organs of perception occurring among plants.

Assuming that the root-tip is the organ for gravi-perception, we are confronted with the remarkable fact that the part of the root which responds to the effect of gravitational stimulation of the root—the elongation region—lies at some distance from the gravi-perceptive root-tip. We are, therefore, bound to infer that the course of events which follows on stimulation of the tip is very similar to that which we have described as following on the stimulation by the pressure of the pinching shoe (p. 201). As a consequence of stimulation, in either case, the organ of perception is excited. Excitation of the perceptive organ results in waves of disturbance or change—nervous impulses—which pass from the perceptive region to the region of reaction (motor region). In the latter region, the nervous impulses set up a disturbance or excitation, as a consequence of which, in the case of the root, growth becomes more rapid on the one side than on the other, and geotropic curvature takes place. We need not, however, suppose that the conduction of the nervous impulse in the root is effected, as it is in most animals, by specialised protoplasmic fibres or "nerves," for, as we have pointed out already, conduction of nervous impulses may take place in animals which have no specialised nerve-fibres. The point of interest is that in the plant-reflex, as in the animal-reflex, perception may be localised and the perceptive region may be separated from the motor region by tissues which respond to the results of stimulation by transmitting a nervous impulse set up as a consequence of the excitation of the perceptive region.

We may recognise, therefore, in the reflexes of plants a series of events similar to those which occur in the reflexes of animals :—perception, excitation, conduction of nervous impulses, excitation and response of the motor apparatus. Of course, there is not necessarily any marked separation between the perceptive region and the reacting region. Such separation we may regard as a sign of specialisation, that is, of increasing efficiency. Thus an organism may be so unspecialised as to have no eyes and yet be capable of perceiving the stimulus of light, and it may, at the same time, be so minute that there is no room for any visible separation between the part which perceives and the part which reacts. In such simple organisms, it may be that the protoplasm which perceives is also that which contracts or reacts in other ways to the consequences of the perception. But, since the work of perception is different in kind from that of reaction, it is probable that, even in very simple organisms, one part of the living protoplasm becomes specially adept at " perceiving " environmental changes and another part becomes specialised to carry out promptly the purposeful reflex which follows as a result of perception, and since, in this simple organism, it is the living substance as a whole which reacts, it must possess means of transmitting the excitation which results from perception of a stimulus to all parts of the protoplasm. We conclude, therefore, that plants respond to stimuli by nervous reflexes, the phases of which consist in perception, excitation, transmission of "nervous impulse," excitation of the motor apparatus and motor-reaction.

It is easy to devise experiments which illustrate further the localisation of perception of stimuli.

211. For this purpose, sow the seeds of Italian millet (Setaria italica) or millet (Sorghum vulgare) in a small box which is kept in a dark place till the seedlings are about half an inch in height. In the meantime, little caps to fit over the tips of the emerging cotyledons are prepared by rolling pieces of tinfoil about one-third of an inch square round the end of a stiff wire. The cotyledons of about half of the seedlings are covered each with one of these elongated, conical caps, the narrow ends of which are so squeezed that the caps are about one-eighth inch long. The remaining seedlings are left uncovered, and the pan is put in a box

open on one side to the light so that the seedlings are illuminated unilaterally.   Examine them from day to day, and
observe that, whereas the uncovered seedlings bend towards
the light, those whose cotyledons have been covered with
tinfoil caps stand bolt upright.   Note also that the positive
phototropic curvature of the uncovered seedling occurs not
in the cotyledon, but in the part (hypocotyl) below the

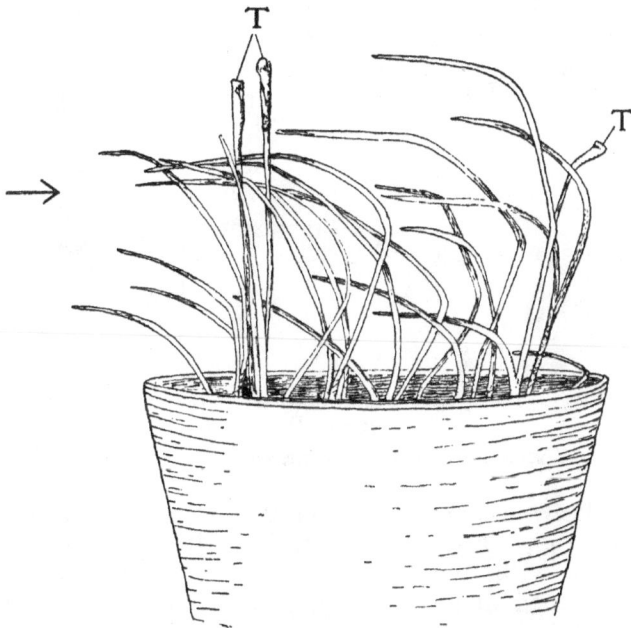

FIG. 28.—SEEDLINGS OF SETARIA ITALICA GROWN IN UNILATERAL LIGHT.

T, tinfoil cap.   The arrow indicates the direction of the light.

(*From a photograph.*)

cotyledon (Fig. 28).   As a handkerchief bound round the
head blindfolds our eyes, so the cap "blindfolds" the
perceptive organs which are situated in the tip of the
cotyledon.

Another method of demonstrating localisation of perceptive organs consists in maintaining the part of the plant
in which these organs occur in such a position that they
are constantly stimulated, and observing that, as a
consequence, the motor region continues to react, with the

result that it becomes more and more curled up into circles or spirals. For this purpose, we make use of the fact that, whereas the motor region for the execution of geotropic and heliotropic response is located in various grass seedlings in the hypocotyl, the perceptive region is located in the cotyledon.

212. Prepare a series of glass tubes, each just wide enough to allow the cotyledon of Setaria or Sorghum to fit into it. Push the tubes horizontally into a heap of moist sand pressed against one side of a biscuit tin. Germinate the seedlings, and when the cotyledons appear, fix in each tube a cotyledon of one of the seedlings and stick the

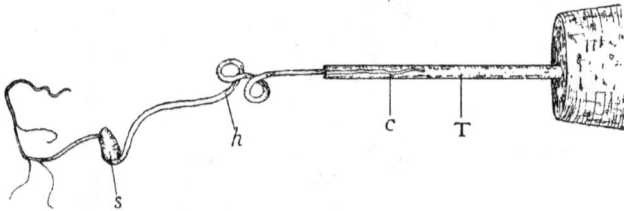

FIG. 29.—SETARIA ITALICA. SEEDLING WITH COTYLEDON FIXED IN A HORIZONTAL POSITION.

c, cotyledon ; h, hypocotyl ; s, seed ; T, glass tube.

other end of the tube into the sand so that it is placed horizontally. Place the box in the dark. After some days, take out a tube with its seedling and observe that the hypocotyl is looped like a wriggling caterpillar. Continue to observe and draw the seedlings at daily intervals, noting that, the longer they are left with their cotyledons horizontal in the tube, the more do the hypocotyls become curved (museum) (Fig. 29). In its enforced horizontal position, the cotyledon is subject constantly to gravitational stimulus. As a consequence, its organs of perception transmit a continuous series of impulses to the motor region, in obedience to which that region executes, and continues to execute, the motor reflex.

213. By a similar experiment, in which the cotyledon of each seedling is imprisoned in a glass tube, demonstrate that, in such grasses as Sorghum, the organs of light-perception are also localised in the cotyledon. When illuminated unilaterally, the hypocotyl bends, as we saw in Exp. 211, so as to bring the cotyledon into the line of light.   If, therefore, light is caused to fall, *e.g.* from the side, on cotyledons of Setaria or Sorghum each imprisoned in a vertical glass tube, then, though in its rest-position with respect to gravity, the cotyledon is stimulated constantly by the light, and transmits impulses to the hypocotyl, which responds by photo-

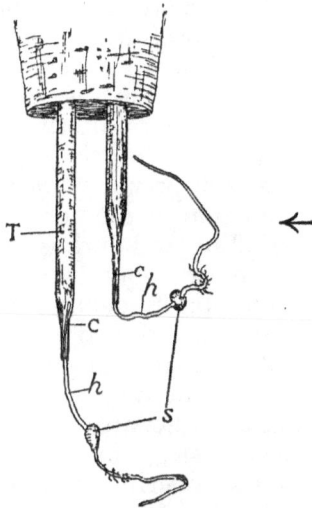

FIG. 30.—SORGHUM VULGARE.   SEEDLINGS WITH COTYLEDON FIXED IN
A VERTICAL POSITION, IN UNILATERAL LIGHT.

c, cotyledon ; *h*, hypocotyl ; *s*, seed ; T, glass tube.   The arrow indicates the direction of the light.

tropistic curvature.   Nothing that the hypocotyl can do, in these exceptional experimental circumstances—except perhaps to extricate, by its wriggling curvatures, the cotyledon from its tube—can bring the cotyledon to its proper rest-position pointing in the direction of the light.   Nevertheless, impelled by the unceasing series of impulses from the ever-misplaced perceptive region, it continues its purposeful but ineffectual motor-response (Fig. 30).

Hitherto, we have considered only tropistic movements, that is, movements which, in response to an external stimulus, bring the plant into a definite position with respect to the line of action of that stimulus. There are, however, many other induced movements, that is, movements which are called forth by external stimuli, which, though purposeful, are not tropistic.

214. Thus, if we bring a pot of tulips into a room we may demonstrate readily, by altering the temperature in the vicinity of the pot, that the flowers respond to a rise of temperature by opening, and to a fall, by closing. Here it is not necessary for the stimulus to proceed from a definite direction for the purposeful movement to occur.

215. In other cases, *e.g.* daisy, etc., similar opening and closing movements may be effected by exposing the plant to light and darkness alternately. Cut-flowers with their stalks in water serve for the experiment, though, since the movement may not proceed quickly, it is well to measure and record the position of the corollas of the strap-shaped florets at the beginning and during each stage of the experiment. Not only flowers but leaves also may respond to changes in illumination by definite movements.

216. Thus, during the day the three leaflets of red clover (Trifolium pratense) are spread out in a fairly horizontal position, but, at night, the two side leaves fold together, and are roofed over by the remaining leaflet. Such "sleep movements" may be induced during the day by covering a plant with a pot or box to exclude the light. After an hour or two the leaves are found to have assumed the nocturnal position. Many other plants, *e.g.* wood sorrel (Oxalis acetosella), the scarlet-runner bean (Phaseolus multiflorus), the sensitive plant (Mimosa pudica), exhibit similar "sleep movements." When we compare these latter, purposeful, reflex movements with tropistic movements, we recognise that both types have this in common, that the movements result in the assumption of a definite position by the responding organ; but, whereas, in the case of tropistic movements, the stimulus acts in a definite direction and induces a change of position which has a definite relation with that direction, in the other type, the stimulus need not proceed from a definite direction, and even if it does, the movement which it induces, although it may be quite definite in character, has no reference to that direction. In tropistic reflexes, a directive stimulus induces a directive response; in the other type of movement, which we may call *nastic,* a diffuse stimulus produces a definite movement, the direction of which the plant determines for itself. Thus we can understand that beside these types of induced movements, called forth by external stimuli, a plant may also exhibit movements which are called forth not by external stimuli but by internal changes, which act as internal stimuli.

217. For example, if we grow a bean on a klinostat in darkness, the hooked plumule remains hooked, in spite of the elimination of light and the equalisation of gravitational stimulation. Such a curvature as that executed by the bean plumule we may call autonomous, in contradistinction to induced movements, that is, those called forth by external stimuli.

To sum up : a plant, like an animal, may respond to stimuli, which may be of external or internal origin. The reactions to internal stimuli are called autonomous, those to external stimuli are called induced (or paratonic). Both internal and external stimuli may set up changes in growth and development—morphogenetic responses—and both may give rise to nastic responses. External stimuli may give rise also to tropistic responses, which differ from induced nastic responses in that their directions bear definite relations to those of the stimuli which evoke them.

The phenomena of irritability exhibited by plants are endless in their variety. Though we have dealt with but few, nevertheless we have studied enough to convince ourselves that a plant possesses, no less than an animal, the power of so adjusting itself to its environment as to perform its work with efficiency, even in a world of constant change. So close is the adjustment that we may well conclude that " life is a relation between the living organism and its environment."

APPENDIX A.

It is assumed that the more important of the laboratory fittings and apparatus mentioned in the following list are available.

Directions for using such apparatus, for bending glass-tubing, boring corks, etc., will be found in text-books of Practical Chemistry (Bibliography, 18).

Beakers (glass: various sizes).
Bell-jars (with or without tubulure: various sizes. See Cloches).
Blow-pipe.
Bunsen burners.
Burette (graduated in c.cm.).
Clamps.
Cloches (cheap bell-jars used in market gardens).
Corks.
Cork-borers.
Cover-glasses (for microscope work).
Desiccator.
Dissecting microscope.
Dissecting needles.
Erlenmeyer flasks.
Evaporating dishes (porcelain).
Files.
Filter funnels (various sizes).
Filter paper.
Filter stands.
Flasks (glass: various sizes).
Glass rods.
Glass slides (for microscope work).
Glass tubing (of various bores, including capillary tubes).
Hot air oven.
Klinostat (p. 208).
Lenses (simple).
Litmus papers.

Micrometer.
Microscope.
Pestle and mortar.
Pipettes (ordinary and graduated, *e.g.* 5 and 10 c.cm.).
Platinum foil.
Platinum wire.
Retort stands.
Rubber corks (various sizes; without holes, with 1, and with 2 holes).
Rubber tubing (various sizes: thin and also thick-walled pressure tubing).
Separating funnels.
Spectroscope.
Spirit lamp.
Stop-watch.
Test tubes (various sizes).
Test tube stands.
Test tube brushes.
Thermometers (graduated in Centigrade scale).
Thistle funnels.
Tools (hammer, saw, &c.).
Tripod stands.
Watch glasses.
Water bath.
Wire (copper, &c.).
Wire gauze.

The glass and other apparatus, as well as the reagents referred to in the text, may be obtained direct from firms making a speciality of the supply of laboratory requisites. The chief firms publish illustrated catalogues which will be found of considerable service in facilitating the selection of apparatus.

LIST OF REAGENTS REQUIRED FOR EXPERIMENTS.

ALBUMIN.—Commercial egg-albumin may be used for tests for proteins.

ALCOHOL (C_2H_5HO).—Pure alcohol (100 %) may be obtained commercially, but it is expensive. Methylated spirit contains about 95-98 % alcohol, and may be used for most purposes instead of absolute alcohol.

AMMONIACAL COPPER SULPHATE (see Exp. 192).—Ammonia added to a solution of copper sulphate produces a precipitate of copper hydroxide which redissolves when more ammonia is added to form a blue solution.

AMMONIUM HYDROXIDE ($NH_4(OH)$).—Strong solution.

AMMONIUM MOLYBDATE (($NH_4)_2MoO_4$).—May be purchased, or prepared as follows: Add 50' gm. molybdic acid or 70 gm. ammonium molybdate to 100 c.c. water, then 100 c.c. of strong ammonia. Stir until dissolved. Pour the solution into 720 c.c. of cold nitric acid (sp.g. 1·20), stirring whilst adding.

BARYTA WATER (BARIUM HYDROXIDE) ($Ba(OH)_2$).—Absorbs carbon dioxide with the formation of barium carbonate ($BaCO_3$) which is precipitated as a white powder. Prepare a saturated solution by adding excess of barium hydroxide to distilled water and filtering. The solution should be used fresh.

BENZOL.—Used as a solvent for various substances, including fats and oils. Should never be used near a naked flame, as it is extremely inflammable.

BIURET REACTION (for proteins).—Place in a test tube a small quantity of the solution to be tested. Add *one or two drops* of a dilute solution of copper sulphate. Add an excess of a solution of caustic soda, or potash ; the solution becomes violet. If ammonia is used instead of soda or potash, the colour is blue instead of violet. Peptones give a rose-red colour with potash or soda and copper sulphate, a reddish-violet with ammonia and copper sulphate. In using this test, special care should be taken to add very little copper sulphate.

CAUSTIC POTASH (POTASSIUM HYDROXIDE)(KOH).—2%, 10%, and 50 % solutions should be made by dissolving 2, 10, and 50 gm. of the solid, each in 100 c.c. of distilled water. They should be kept in glass-stoppered bottles, and care taken to avoid spilling the potash on the hands or clothes.

CAUSTIC SODA (SODIUM HYDROXIDE) (NaOH).—Solutions of known percentages made in the same way as those of caustic potash.

K.P. P

CHLORAL HYDRATE, IODINE SOLUTION.—Dissolve 8 parts of
chloral hydrate in 5 parts of water ; add crystals of iodine
which dissolve slowly and colour the solution. Chloral hydrate
causes starch grains to swell, thus rendering them more easy
of detection while still included in the chloroplasts. The
material to be tested should, if green, be decolourised by
placing in methylated spirit, and then laid in the solution
for twelve to twenty-four hours.

COCO-BUTTER.—(See Wax.)

COPPER SULPHATE (CuSO₄).—10 % solution in water.

CORROSIVE SUBLIMATE (MERCURIC CHLORIDE) (HgCl₂).—A very
dilute solution (1 %) in water may be used for antiseptic
purposes as described. Being extremely poisonous, bottles
containing it should be marked " Poison," and kept in a safe
place.

DIASTASE.—Various commercial preparations of diastase may be
obtained, *e.g. malt diastase*, prepared from germinating
barley, etc.; *ptyalin*, prepared from the salivary glands of
animals. *Liquor pancreatici* contains diastatic and also pro-
teolytic enzymes.

DISTILLED WATER.—All reagents should be made up with distilled water unless the contrary is stated. It may be obtained
in large vessels (carboys) from any firm supplying laboratory
reagents, or, if a *still* is available, it may be distilled in the
laboratory.

EUCALYPTUS OIL.—Antiseptic: prevents the development of
micro-organisms in fluids, such as solutions of proteins, etc.

FEHLING'S SOLUTION.—A reagent for the identification of
reducing sugars : it may be obtained commercially or prepared
as follows: *Solution A*—Dissolve 36·64 gm. copper sulphate in
500 c.c. water. *Solution B*—Dissolve 173 gm. sodic-potassium
tartrate (Rochelle salt) in 100 c.c. caustic soda (sp. grav. 1·34)
and dilute with water to 500 c.c. Both solutions should be
kept in well-stoppered bottles, and mixed in equal quantities
for use.

HYDROCHLORIC ACID (HCl).—Concentrated and 10% solutions in
water. This and the other acids required should be kept in
well-stoppered and properly-labelled bottles.

IODINE SOLUTION.—The solution of iodine in common use in the
laboratory for the identification of starch is prepared as
follows: Make a strong solution of potassium iodide in distilled water. Add to this solution some crystals of iodine and
set on one side for some hours, shaking occasionally: dilute
with distilled water to the colour of brown sherry. The

reagent may also be made by diluting the *liquor iodi* of the Pharmacopœia. This solution stains (1) protein substances, brown; (2) cellulose, faintly yellow; (3) starch, blue; (4) succeeded by strong sulphuric acid (2 parts concentrated acid to 1 part water) it stains cellulose blue.

IODINE AND SULPHURIC ACID TEST FOR CELLULOSE.—Soak a small piece of the material to be tested in iodine solution. Place a *small* drop of concentrated sulphuric acid on a glass slide, dilute with rather less than half the quantity of water and transfer the material to this; before microscopic examination put a cover glass over the object. Under the *low* power of the microscope it can be seen that the cell-walls swell, lose their sharp contour, and assume a blue colour, although the colour is often not uniform. Great care should be taken that the acid does not get upon any part of the microscope. A very small drop is sufficient, and the slide and glass rod should be washed after use.

LEAD ACETATE. A saturated watery solution.—Organic matter containing sulphur liberates sulphuretted hydrogen (H_2S) when heated with soda-lime. The H_2S reacts with lead acetate to form lead sulphide. Strips of filter paper dipped in lead acetate solution and exposed to the action of sulphuretted hydrogen turn black owing to the formation of the sulphide.

LIME-WATER (CALCIUM HYDROXIDE) ($Ca(OH)_2$).—At the ordinary temperature, 100 parts of water dissolve ·14 of slaked lime. This weak solution, known as *lime-water*, has an alkaline reaction, and absorbs carbon dioxide, forming chalk (calcium carbonate, $CaCO_3$), which is precipitated as a white powder. It should be freshly prepared. (See Baryta water.)

LIQUOR PANCREATICI.—For the investigation of the digestive action of pancreatic juice, an artificially prepared extract is usually employed which can be obtained commercially under the above name. It may also be obtained in the form of a powder, under the name of *pancreatin.*

LITMUS.—Some vegetable colours are affected by the addition of certain chemicals. The litmus test papers in common use are made by soaking paper in a solution of litmus. This solution in water or alcohol is at once turned *red* by acids, and the reddened litmus is sensitive to alkalies, which give the original blue colour. Similarly, turmeric which is normally yellow, is turned brown by alkalies, the yellow colour being restored by acids.

MAGNESIUM SULPHATE ($MgSO_4$, $7H_2O$). EPSOM SALTS.—A saturated solution in water is used for the precipitation of certain proteins.

METHYLENE BLUE.—A blue dye soluble in water, used in the laboratory for staining various tissues in the plant and animal body. In very weak solution it is taken up by both plant and animal cells, without interference with their vital activities. If presented to a plant, *e.g.* a water-plant, in extremely dilute solution it is taken up by the cells, passing through the plasmatic membrane and staining the cell-sap blue.

MILLON'S REAGENT.—It is best to purchase this reagent, which consists of a mixture of mercurous and mercuric nitrates, made up ready for use. Place in a test-tube a little of the protein solution to be tested. Add a few drops of the reagent. A white precipitate is formed, which, on heating, *becomes brick-red.* The latter part of the reaction is the essential one as a test for proteid.

NITRIC ACID (HNO_3).—Concentrated and 10 % solutions in water. (See note on Hydrochloric Acid.)

OSMIC ACID.—A ·1 to 1 % solution in water stains fats black. It is an expensive reagent, and should be kept in a well-stoppered bottle in the dark. Do not let the vapour get into the eyes as it sets up irritation.

PARAFFIN WAX.—Paraffins of different degrees of hardness and of different melting points may be obtained from dealers in microscopical requisites. Soft paraffin, melting point 36°, and a harder paraffin, melting point 54°-58°, should be obtained, and, if required, paraffins of intermediate melting points may be prepared by mixing these in varying proportions. To ascertain the melting point, melt a little wax in a dish, take up a drop on the end of the bulb of a thermometer and read off the temperature at which the paraffin begins to solidify in the air.

PASTEUR'S NUTRIENT SOLUTION.—The nutrient solution first used by Pasteur for the cultivation of micro-organisms is prepared as follows :—

> 100 c.c. water,
> 10 gm. cane-sugar,
> ·1 „ ammonium tartrate,

and the ash obtained by incinerating 1 gm. of yeast.

PEPSIN.—An extract obtained from the walls of the stomach of pig or other animal. Obtainable in commerce. Contains a protease which hydrolyses proteins to peptones.

PEPTONE.—Peptones are substances formed by the partial hydrolysis of proteins. Artificially prepared peptone may be obtained commercially.

PHENOLPHTHALEIN.—One of the substances which, like litmus, are known to chemists as indicators, because their colour changes according as the medium is acid or alkaline. Phenolphthalein is colourless in neutral or acid solutions, red in alkaline solutions. When a base such as caustic soda or potash is added gradually to an acid solution containing the indicator the acid is just neutralised, and the slightest excess of alkali is then indicated by a change of colour in the solution.

POTASSIUM BICHROMATE ($K_2Cr_2O_7$).—10 % solution (saturated). The orange-red solution is extremely poisonous.

POTASSIUM IODIDE (KI).—Strong solution in water. Used in the preparation of iodine solution. (See Iodine.)

POTASSIUM NITRATE (KNO_3).—10 %, 5 %, 2·5 % solutions in water from which solutions of intermediate strength may be prepared. Used for estimating the osmotic pressure of cell-sap.

PROTEINS.—Solutions of egg-albumin and of plant-proteins may be prepared as described in the text. Pure egg-albumin may be obtained commercially.

SODA-LIME.—Granulated. Used for the absorption of CO_2 from air.

SODIUM CHLORIDE (COMMON SALT) (NaCl).—A 10 % solution should be made, and diluted to the strength desired for experiments in plasmolysis, etc.

STARCH ($(C_6H_{10}O_5)_x$).—Pure starch may be obtained commercially. —The following reactions are characteristic of starch grains :—
    (i) Colour blue with iodine in the presence of water.
    (ii) Swell in a solution of caustic potash or chloral hydrate.
    (iii) Swell and colour blue with a solution of iodine in chloral hydrate.
    (iv) Swell in water above 65° C.
    (v) Swell in dilute sulphuric acid.

SUGARS.—Pure glucose, sucrose, and other sugars may be obtained commercially.

SULPHURIC ACID (H_2SO_4).—Concentrated and 10 % solutions in water. (See note on Hydrochloric Acid.)

THYMOL.—Antiseptic serving to prevent the growth of micro-organisms in putrescible fluids.

VASELINE.—Used in making wax mixture, etc.

WAX.—A wax mixture suitable for making joints air-tight, etc., is made as follows :—

> Resin,   15 parts.
> Beeswax,   35   „
> Vaseline,   50   „

The wax and vaseline are melted together, the resin is powdered and added gradually to the hot mixture, stirring meanwhile. The hardness of the mixture may be controlled by varying the proportions of the beeswax and vaseline.

(Coco-butter may be used, *e.g.* in Exp. 159.)

WATER CULTURES.

NORMAL SOLUTION.—

> Potassium nitrate,   1·0 gm.
> Ferrous phosphate,   ·5   „
> Calcium sulphate,   ·25 „
> Magnesium sulphate,   ·25 „
> Distilled water,   1-2 litres.

LACKING CALCIUM.—

> Potassium nitrate,   1·0 gm.
> Ferrous phosphate,   ·5   „
> Magnesium sulphate,   ·5   „
> Distilled water,   1-2 litres.

LACKING POTASSIUM.—

> Calcium nitrate,   1·0 gm.
> Ferrous phosphate,   ·5   „
> Calcium sulphate,   ·25 „
> Magnesium sulphate,   ·25 „
> Distilled water,   1-2 litres.

LACKING NITRATES.—

> Potassium chloride,   1·0 gm.
> Ferrous phosphate,   ·5   „
> Calcium sulphate,   ·25 „
> Magnesium sulphate,   ·25 „
> Distilled water,   1-2 litres.

LACKING PHOSPHATES.—

> Potassium nitrate,   1·0 gm.
> Ferric chloride, 1 or 2 drops of a watery solution.
> Calcium sulphate,   ·25 gm.
> Magnesium sulphate,   ·25 „
> Distilled water,   1-2 litres.

LACKING MAGNESIUM.—

Potassium nitrate,	1·0	gm.
Ferrous phosphate,	·5	,,
Calcium sulphate,	·5	,,
Distilled water,	1-2	litres.

LACKING IRON.—

Potassium nitrate,	1·0	gm.
Calcium phosphate,	·5	,,
Calcium sulphate,	·25	,,
Magnesium sulphate,	·25	,,
Distilled water,	1-2	litres.

Certain precautions are necessary for success in growing water-cultures.

The bottles must be carefully cleaned, and pure distilled water should be used for making up the solutions.

The chemicals used must be pure, and, if a stock is made up for refilling the bottles, the solutions should be shaken up thoroughly before using, since some of the salts are only slightly soluble, and may precipitate on standing.

Stock bottles should be well stoppered and kept in the dark.

In addition to these precautions it is necessary to aerate the bottles daily.

This is most easily effected by fixing an open glass tube in a second hole in the cork of the culture-vessel and blowing air through it daily, by means of an india-rubber ball, such as is used for spraying scent, etc.

The solutions should be changed every ten days.

The roots of the plants should be kept in darkness, either by sinking the culture-vessels in a box of sawdust or by wrapping them in black cloth.

TO CLEAN THE BOTTLES.—Wash thoroughly with nitric acid. Rinse repeatedly with tap water and then with distilled water. Finally, they may be washed out with a weak solution of corrosive sublimate (·1 % sol.) and again thoroughly rinsed with distilled water.

A simple method for growing water-cultures of small seedlings, *e.g.* cress, has been described by Osterhout as follows :—

Fold a strip of filter paper lengthwise, as shown in fig. 31, and place it round the inner side of the upper part of the top of an ordinary tumbler. The ends of the strips may be fastened with paper clips.

Having previously paraffined the inside of the tumblers, fill them with the culture solution and place the seeds in the paper troughs in contact with the solution.

Water should be added occasionally to replace that lost by evaporation.

XANTHOPROTEIC REACTION (FOR PROTEINS).—Place in a test tube a small quantity of the solution to be tested. Add a few drops of strong nitric acid. With certain proteins, a white precipi-

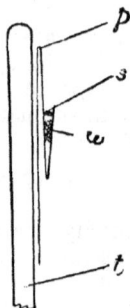

FIG. 31.—WATER CULTURE. A SIMPLE METHOD. (See p. 231.)
DIAGRAMMATIC VIEW OF APPARATUS.

t, tumbler ; *p*, filter paper ; *s*, seeds ; *w*, water level.

tate is produced, which turns yellow on boiling. With other proteins no precipitate is formed, but the *liquid* turns yellow on boiling. Add a few drops of ammonia ; the yellow colour changes to *orange*. The change in colour on adding ammonia is the essential part of the reaction, and is a very delicate test for proteins.

B.

APPARATUS, MATERIALS, ETC.

APPARATUS FOR PREPARATION OF GASES.—A description of the mode of preparation and of the necessary apparatus for preparing nitrogen, hydrogen, carbon dioxide, etc., may be found in any text-book of practical chemistry (Bibliography, 18).

APPARATUS FOR FREEING WATER FROM ITS DISSOLVED GASES (see p. 66).—Close the opening of a 1000 c.c. flask with a well-fitting rubber cork provided with one small hole. Remove the cork and pass through the hole a piece of narrow-bore glass tubing about 6 ft. long, so that its lower end is about half an inch below the lower surface of the cork. Fill the flask nearly full with water. Boil for about 2 hours. Whilst the water is boiling, push the cork into the flask so that the tube dips just beneath the surface of the water. Remove the source of heat and allow the flask to cool. The cooling may be hastened by plunging the flask, after the temperature of the water has fallen somewhat, into a pail of cold water. The long narrow tube fixed in the cork, being practically full of water, prevents the latter from taking up more than small quantities of air. The water should be used, *e.g.* in Exp. 58, as soon as it is cold.

BALANCES.—Balances for use in experiments which involve weighing can be obtained from any of the firms which supply scientific and chemical apparatus.

For rough estimations a Beranger balance is useful. Balances of this type may be obtained in various sizes to weigh from 1 to 5 kg., at prices from 21s. to 30s.

For more accurate work, a chemical balance is necessary. Such a balance, suitable for students' work, adapted to carry a load of 100 gr., and sensitive to ½ mg., can be obtained for about £2 7s.

COCO FIBRE.—This material may be obtained from any nurseryman. It may with advantage be used instead of soil for germinating seeds, etc. It is more satisfactory for this purpose than sawdust, in which the roots of seedlings often become unhealthy.

DEWAR FLASK.—In addition to the ordinary type of thermo-flask, others may be obtained of the shape of an ordinary flask, with a narrow neck and mounted on a wooden base. These are suitable for such experiments as those described on p. 74. They should be tested before using for experimental work by filling with warm water of known temperature, and determining the rate of cooling.

DARK BOX.—A light-proof box, as shown in fig. 12, can be made by any carpenter. A useful size is as follows:—Height, 1 ft. 8 in.; width, 1 ft. 2 in.; length, 1 ft. 8 in.; shutter, 1 ft. square. The lid and shutter should fit closely so as to completely exclude the light, and the latter may be replaced by screens of coloured glass, etc., when required for special experiments. An aerating-tube is indicated in fig. 12 (T).

DIFFUSION SHELLS.—See parchment membrane.

DOUBLE BELL JARS.—Double bell jars, for holding coloured solutions as described in Exp. 192, may be obtained in sizes varying from 4¾ in. to 7 in. diameter. Being portable, they are useful for making observations on the effect of coloured light on plants growing in the open, but they have the disadvantage of being rather expensive.

ERLENMAYER FLASK.—These flasks (see Catalogues of apparatus, etc.) are obtainable in various sizes, with or without a side tube.

ESTIMATION OF AREA OF A GIVEN NUMBER OF LEAVES.—A simple means of calculating the area of a number of leaves is as follows :—
Draw on scribbling paper outlines of the leaves whose area is to be measured; cut out the paper patterns with sharp scissors. Weigh a known area—say a single sheet—of the same paper, and record the weight ; weigh the paper leaves, and from these data determine, by simple calculation, the area of the leaves.

GELATINE PLATES.—Thin sheets of coloured gelatine, which may be used as described in Exp. 193, may be obtained from dealers in laboratory requisites (see Catalogues, e.g. Baker's, London).

GERMINATORS.—Small numbers of seeds may be readily germinated in saucers or dishes with layers of blotting-paper, as described in the text.
Larger germinators of different patterns for seed-testing experiments, etc., are supplied by makers of scientific requisites, whose catalogues should be consulted.
Germinating boxes with sloping glass fronts (seedling observation boxes) as shown in fig. 32 can be made by a carpenter. A

useful size is as follows :—Length, 9 in. ; depth, 5 in. ; width at top, 6 in. ; width at bottom, 4½ in.

INDIA-INK MARKS ON SEEDLINGS.—Bend a piece of stout wire about 6 in. long, make a small loop at each end and tie a fine silk thread to each loop so that it is tightly stretched ; moisten the thread with waterproof india-ink ; lay the seedling to be marked on white paper or a plate of cork alongside a scale or ruler, and make a series of marks at approximately equal intervals by pressing the inked thread gently on the surface of the stem or root.

Whilst the ink is drying measure the distances between the marks ; dip the root in water and replace the marked seedling in a germinator.

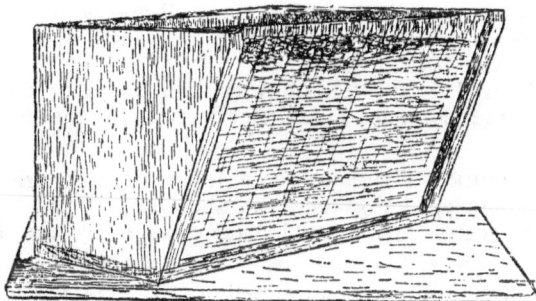

FIG. 32.—OBSERVATION BOX FOR SEEDLINGS. (See p. 234.)

MEASURING JARS.—Graduated measuring jars of varying capacities are useful for making up solutions.

For measuring small amounts it is best to use a graduated pipette. A burette is often convenient when the quantity of liquid to be added is not definitely known beforehand. For method of using pipettes, etc., see a text-book of practical chemistry (Bibliography, 18).

OBSERVATION BOX.—See Germinators.

PARCHMENT MEMBRANE.—Obtainable in sheets or in the form or a tube. Diffusion shells of test-tube shape are also obtainable and are useful for experiments on osmosis. The solution is poured into one of these little parchment tubes, which is then suspended by a glass rod as described in Exp. 17.

Parchment tubing when not in use is best kept in a weak solution of formalin (1 %).

PLATINUM WIRE.—A piece of platinum wire about 2 inches long is fused into the end of a piece of glass tubing which acts as a holder. The free end of the wire is then bent into a loop and is ready for use.

RUBBER TUBING.—Rubber tubing of different bores—from $\frac{1}{16}$ in. to 1 in. may be purchased by the foot. A supply of the most useful sizes should be kept in stock. Stronger tubing (pressure tubing) is advisable for certain experiments.

SAND CULTURES.—Procure clean quartz sand, such as the "silver sand" sold for horticultural purposes. Wash it several times with weak sulphuric acid, rinse repeatedly with tap water and finally with distilled water.

Place the sand so prepared in clean, new flower-pots, and sow the seeds in the usual way. Shade until the seedlings appear, if necessary watering with distilled water. Water with the solutions prepared for water-cultures, and record the growth of the seedlings at short intervals.

Sand prepared as directed may contain traces of soluble iron salts, and is, therefore, not suitable for determining the effect of lack of iron.

TEMPERATURE.—There are two different thermometric scales in use in this country, Centigrade and Fahrenheit.

The two scales are mutually convertible by the following formulae :—

Centigrade to Fahrenheit :—$(C.° \times \frac{9}{5})+32.$
Fahrenheit to Centigrade :—$(F.° - 32) \times \frac{5}{9}.$

WEIGHTS AND MEASURES.

CONVERSION TABLES.

I.—Length.

Metres to yards	Multiply by	1·094
„ ft.	„	3·281
Centimetres to inches	„	·3937
Yards to metres	„	·9144
Feet to metres	„	·3048
Inches to centimetres	„	2·540

II.—Surface.

Sq. metres to sq. yards	Multiply by	1·196
„ „ „ ft.	„	10·76
„ centimetres to sq. inches	„	·1550
„ yards to sq. metres	„	·8361
„ feet to sq. metres	„	·0929
„ inches to sq. centimetres	„	6·452

III.—Volume.

Cu. metres to cu. ft.	- - -	Multiply by	35·32
„ centimetres to cu. inches	- -	„	·06102
Litres to gallons	- - - -	„	·2200
Litres of water to pounds	- -	„	2·205
Cu. feet to cu. metres	- - -	„	·02832
„ inches to cu. centimetres (c.cm.) -		„	16·39
Gallons to litres	- - - -	„	4·546
Lbs. of water to litres	- - -	„	·4538

IV.—Weight.

Grammes (gm.) to ounces	- -	Multiply by	·03527
„ „ grains -	- - -	„	15·43
Ounces to grammes -	- - -	„	28·35
Grains to grammes -	- - -	„	·06478

WIRE.—Fine copper wire is useful for making joints secure, etc. The fine iron wire used by florists for button-holes, etc., is also useful for this purpose, and can be obtained on reels from florists or dealers in horticultural requisites.

C.

LIST OF PLANTS USED IN EXPERIMENTS.

(Unless otherwise stated, plants are to be used in a living and fresh condition.)

ACER SP. (Maple).—Garden varieties. Variegated leaves for starch. Summer.

ACER PSEUDO-PLATANUS (Sycamore).—Fruits. Germination and museum. Autumn.

ACONITUM NAPELLUS (Monkshood).—Garden perennial. Leaves for hydathodes. Summer.

ÆSCULUS HIPPOCASTANUM (Horse-chestnut).—Leaves for leaf-fall. Autumn. Fresh or in alcohol. Fruits and seeds, for germination and museum.

AGAVE SP.—Greenhouse and half-hardy garden perennials. Leaves, for xerophytic characters ; epidermis and stomata.

ALISMA PLANTAGO (Water plantain).—Common native aquatic plant. Leaves, for stomata. Summer. Seeds, for germination under water. August.

ALLIUM CEPA (Onion).—Bulbs can be obtained at any season, and grown in hyacinth glasses.

ALOË SP.—Greenhouse and half-hardy perennials. Leaves, for xerophytic characters ; epidermis and stomata.

AMPELOPSIS HEDERACEA (Virginia creeper).—Tendrils. Summer. Leaves in autumn, for chromatophores.

ANTIRRHINUM MAJUS (Snapdragon).—Hardy garden perennial. Flower, for coloured cell-sap. Summer.

ARUM MACULATUM (Cuckoo-pint).—Leaves.

AVENA SATIVA (Oat).—Grains, for germination and seedlings. Starch grains.

BARLEY.—(See Hordeum.)

BEAN.—(See Phaseolus and Vicia.)

BELLIS PERENNIS (Daisy).—Flowers, for opening and closing movements.

BERTHOLLETIA EXCELSA (Brazil nut).—Seed, for aleurone grains and oil.

BETA VULGARIS (Beet).—Root, for coloured cell-sap. Sugar,

BRASSICA ALBA OR BRASSICA NIGRA (Mustard).— Seed, for germination and seedlings.

BRASSICA RAPA (Turnip).—Seeds, for germination and seedlings. Root for sugar.

BRAZIL NUT.—(See Bertholletia.)

CALLITRICHE VERNA (Water Starwort).—Common aquatic in ditches and ponds. Leaves, for hydathodes and stomata. Summer.

CALLUNA VULGARIS (Ling).—A common heath plant. For xerophytic characters.

CALTHA PALUSTRIS (Marsh Marigold). Common native in wet ground, by ditches, etc. Leaves.

CARROT.—(See Daucus carota.)

CHLAMYDOMONAS SP.—A unicellular, motile green plant, common in stagnant water. For microscopic examination.

CLOVER.—(See Trifolium.)

COWSLIP.—(See Primula officinalis.)

CUCURBITA PEPO (Vegetable marrow).—Seed, for germination and seedling stages.

DAISY.—(See Bellis perennis.)

DANDELION.—(See Taraxacum.)

DATE.—(See Phœnix.)

DAUCUS CAROTA (Carrot).—Root, for sugar and chromatophores.

DELPHINIUM SP. (Larkspur).—Garden annuals or perennials. Leaves, for hydathodes. Summer.

DROSERA ROTUNDIFOLIA (Sundew).—Native insectivorous plant common on wet heaths.

ECHEVERIA SP.—Succulent half-hardy garden perennials; often used for bedding. For xerophytic characters.

ELDER.—(See Sambucus.)

ELODEA CANADENSIS (American waterweed).—Common floating aquatic in ponds, streams and ditches. Shoots, for experiments in photosynthesis.

ERICA CINEREA (Heath).—Common native on heaths and moors. For xerophytic characters.

ESCHSCHOLTZIA CALIFORNICA (Californian poppy).—Hardy garden annual. Seedlings for sand cultures.

EUGLENA VIRIDIS.—A unicellular green animal, common in stagnant water. For comparison with Chlamydomonas.

FAGUS SYLVATICA (Beech).—Leaves. Summer. Sun and shade leaves.

FICUS ELASTICA (Indiarubber plant).—Commonly grown as a pot-plant in greenhouses and dwelling-rooms. Leaves, for xerophytic characters.

FLAX.—(See Linum.)

FRAXINUS EXCELSIOR (Ash).—Fruits and seeds. Germination and seedling stages.

FROG-BIT.—(See Hydrocharis.)

FUCHSIA SP.—Greenhouse and half-hardy garden perennials. Leaves, for hydathodes. Pot-plants for root-pressure experiments.

GERANIUM.—(See Pelargonium.)

HEDERA HELIX (Ivy).—Climbing stems.

HELIANTHUS ANNUUS (Sunflower).—Common garden annual. Seeds. Germination and seedlings. Shoots. Summer.

HELIANTHUS TUBEROSUS (Jerusalem artichoke).—Garden perennial. Young pot-plants. Summer.

HELLEBORUS NIGER (Christmas Rose). — Garden perennial. Leaves, for hydathodes.

HORDEUM VULGARE (Barley).—Seed, for germination and seedlings. Starch grains.

HYACINTHUS ORIENTALIS (Hyacinth).—Bulbs for root-system. Leaves, for epidermis and stomata.

HYDROCHARIS MORSUS-RANAE (Frog-bit). — Common native, floating aquatic in ditches. Roots and root-hairs.

IRIS GERMANICA.—Hardy garden perennial. Rhizome, for starch grains.

IVY.—(See Hedera.)

LATHYRUS SP.—Climbing stems ; tendrils.

LILIUM SP. (Lily).—Leaves, for stomata and for starch.

LIME.—(See Tilia).

LINUM USITATISSIMUM (Flax).—Seeds, for oily reserves. Young plants for photosynthesis experiments.

LIVERWORTS.—Any common species. Thallus, for starch.

LUPINUS SP. (Lupine).—Garden annuals and perennials. Seed ; dry, for aleurone grains.

MAIZE.—(See Zea Mays.)

MAPLE.—(See Acer.)

MIMOSA PUDICA (Sensitive plant).—Greenhouse annual. Seedlings and young plants, for movements.

MOSSES.—Any common species. Leaves, for starch.

NARCISSUS SP.—Garden perennials. Flower-peduncle, for reaction to gravity.

OAT.—(See Avena.)

ONION.—(See Allium.)

ORNITHOGALUM UMBELLATUM (Star of Bethlehem).—Hardy garden perennial. Seeds; dry, for endosperm.

ORYZA SATIVA (Rice).—Seed, for germination under water.

PELARGONIUM ZONALE (Geranium).—Greenhouse and half-hardy perennials, often grown as pot-plants. Leaves, for hydathodes.

PHASEOLUS MULTIFLORUS (Scarlet-runner bean).—Seed, for germination and seedlings.

PHASEOLUS VULGARIS (Kidney bean).—Seed, for germination and seedlings. Leaves, for pulvini.

PHŒNIX DACTYLIFERA (Date palm).—Seed, for germination and seedlings.

PINUS SYLVESTRIS (Scotch fir).—Leaves, for xerophytic characters. Seedlings and young plants.

PISUM SATIVUM (Pea).—Seed, for germination and seedlings. Starch and aleurone grains.

POTAMOGETON NATANS (Broad-leaved pond-weed).—Leaves, for stomata.

PRIMULA OFFICINALIS (Cowslip).—Flower-peduncle.

PRIMULA SINENSIS (Chinese primula).—Greenhouse perennials. Leaves and petioles, for coloured sap and stomata.

RANUNCULUS AQUATILIS (Water Crowfoot).—A common aquatic in ditches and streams. Leaves.

RAPHANUS RAPHANISTRUM (Radish).—Seed. Germination and seedlings.

RHODODENDRON SP. (including Azalea).—Garden and greenhouse shrubs. Chalk-shy (Calciphobe).

RIBES SANGUINEUM (Flowering Currant).—Common garden shrub. Opening buds, April.

RICE.—(See Oryza sativa.)

RICINUS COMMUNIS (Castor-oil plant).—Half-hardy garden perennial. Seed, for germination and seedlings. Aleurone grains and oil.

SACCHAROMYCES CEREVISIÆ (Yeast).—Cell structure.

SAMBUCUS NIGRA (Elder).—Twigs of various ages.

SAXIFRAGA SP. (Saxifrage).—Common garden perennials. Leaves, for hydathodes and chalk-glands. Fresh and in alcohol.

SCILLA SP.—Garden perennials. Leaves, for starch.

SCOTCH FIR—(See Pinus sylvestris.)

SEDUM SP.—Common hardy perennials. Xerophytic characters. Leaves, for epidermis and stomata.

SETARIA ITALICA (Italian Millet).—A half-hardy grass. Seedlings, for tropistic curvatures.

SMUT.—(See Ustilago.)

SOLANUM TUBEROSUM (Potato).—Tuber.

SORGHUM VULGARE (Millet).—A half-hardy grass. Seedlings, for tropistic curvatures.

SPARMANNIA AFRICANA.—Greenhouse perennial. Shoots, for root-pressure experiments.

SPHAGNUM SP.—The characteristic moss of peat-bogs.

SUNDEW.—(See Drosera.)

SUNFLOWER.—(See Helianthus annuus.)

TARAXACUM OFFICINALE (Dandelion).—Root, for root-cuttings. Flower-scape, for experiments on osmotic pressure.

TILIA EUROPÆA (Lime).—Branches, for transpiration experiments.

TRADESCANTIA VIRGINIANA (Spiderwort).—Hardy garden perennial. Flower. Summer. Staminal hairs for plasmolysis experiments.

TRADESCANTIA ZEBRINA.—Common greenhouse perennial. Leaves, for epidermis and stomata. Cuttings.

TRIFOLIUM PRATENSE (Red clover).—Growing plants for experiments on photosynthesis and sleep-movements.

TRITICUM VULGARE (Wheat).—Seed, for germination and seedlings. Starch.

TROPÆOLUM MAJUS (Nasturtium, Indian cress). — Common garden annuals. Flower, for chromatophores. Leaves, for hydathodes.

TULIPA SP. (Tulip).—Garden perennials. Bulb. Leaves, or epidermis and stomata.

ULEX EUROPÆUS (Furze, Gorse).—Seed, for germination and seedlings. Shoots, for xerophytic characters.

USTILAGO SP. (Smut).—A parasitic fungus, attacking cereals, etc. Spores.

VICIA SEPIUM (Bush vetch).—A common hedgerow plant. Leaves, for hydathodes. Tendrils.

WHEAT.—(See Triticum.)

ZEA MAYS (Maize).—Seed, for germination and seedlings.

BIBLIOGRAPHY.

GENERAL BOTANY.

1. FARMER, J. BRETLAND, *The Book of Nature Study.* Edited by J. Bretland Farmer, M.A., D.Sc. (Oxon.), F.R.S., assisted by a staff of specialists. 5 vols. Fully illustrated. London: The Caxton Publishing Company.

2. BOWER AND GWYNNE-VAUGHAN, *Practical Botany for Beginners.* By F. O. Bower, Sc.D., F.R.S., and D. T. Gwynne-Vaughan, M.A. London: Macmillan & Co., Ltd.

3. GROOM, P., *Elementary Botany.* By Percy Groom, M.A. (Cantab. and Oxon.), D.Sc. (Oxon.), F.L.S., F.R.H.S. Illustrated. London: G. Bell & Sons, Ltd.

4. GROOM, P., *Trees and Their Life Histories.* By Percy Groom, M.A. (Cantab. and Oxon.), D.Sc. (Oxon.), B.Sc. (Birm.), F.L.S., F.R.H.S. Illustrated. Cassell & Company, Ltd.

5. PERCIVAL, J., *Agricultural Botany: Theoretical and Practical.* By John Percival, M.A., F.L.S., F.C.S. London: Duckworth & Co.

6. SCOTT, D. H., *An Introduction to Structural Botany.* Part I., Flowering Plants; Part II., Flowerless Plants. By Dukinfield Henry Scott, M.A., Ph.D., F.R.S., F.L.S., F.G.S. Illustrated. London: Adam and Charles Black.

PLANT PHYSIOLOGY.

7. DARWIN AND ACTON, *Practical Physiology of Plants.* By Francis Darwin, M.A., F.R.S., and Hamilton Acton, M.A. Illustrated. Cambridge University Press.

8. PFEFFER, W., *The Physiology of Plants.* A Treatise upon the Metabolism and Sources of Energy in Plants. By Dr. W. Pfeffer. Translated and edited by Alfred J. Ewart, D.Sc., Ph.D., F.L.S. Illustrated. Oxford: Clarendon Press.

9. SCHIMPER, A. F. W., *Plant Geography upon a Physiological Basis*. By Dr. A. F. W. Schimper. English translation by W. R. Fisher, B.A., revised and edited by Percy Groom, M.A., D.Sc., F.L.S., and Isaac Bayley Balfour, M.A., M.D., F.R.S. Illustrated. Oxford: Clarendon Press.

GENERAL SCIENCE.

10. COHEN, J. B., *Theoretical Organic Chemistry*. By Julius B. Cohen, Ph.D. London: Macmillan & Co., Ltd.

11. COHEN, J. B., *Organic Chemistry for Advanced Students*. By Julius B. Cohen, Ph.D., B.Sc. London: Edward Arnold.

12. GEIKIE, A., *Outlines of Field Geology*. By Sir Archibald Geikie, F.R.S. London: Macmillan & Co., Ltd.

13. GEIKIE, A., *Text-book of Geology*. By Sir Archibald Geikie, F.R.S. London: Macmillan & Co., Ltd.

14. GREGORY AND HADLEY, *A Class-book of Physics*. By R. A. Gregory and H. E. Hadley, B.Sc. London: Macmillan & Co., Ltd.

15. JONES, D. E., *Elementary Lessons in Heat, Light, and Sound*. By D. E. Jones, B.Sc. London: Macmillan & Co., Ltd.

16. HALL, A. D., *The Soil*. An introduction to the Scientific Study of the Growth of Crops. By A. D. Hall, M.A. (Oxon.). London: John Murray.

17. M'CONNELL, P., *The Elements of Agricultural Geology*. By Primrose M'Connell, B.Sc., F.G.S. London: Crosby, Lockwood & Son.

18. NEWTH, G. S., *Elementary Practical Chemistry*. By G. S. Newth, F.I.C., F.C.S. Illustrated. London: Longmans, Green & Co.

19. SMITH, A., *Introduction to General Inorganic Chemistry*. By Alexander Smith, B.Sc., Ph.D., F.R.S.E. London: G. Bell & Sons, Ltd.

INDEX.